||| | ||||||||| ||| | ||| | |||||| |||
I0034474

Cadaver Dissection with Clinical Applications

Seth Gardner D.C., M.S.

Associate Professor of Anatomy
Lake Erie College of Osteopathic Medicine
(LECOM)
Bradenton, Florida

(CRC) CRC Press
Taylor & Francis Group
Boca Raton London New York

CRC Press is an imprint of the
Taylor & Francis Group, an **informa** business

A SCIENCE PUBLISHERS BOOK

Cover Image taken by Seth Gardner.
Book illustrations were created by Salvadore Raga specifically for this dissection manual.

First edition published 2024
by CRC Press
2385 NW Executive Center Drive, Suite 320, Boca Raton FL 33431

and by CRC Press
4 Park Square, Milton Park, Abingdon, Oxon, OX14 4RN

© 2024 Seth Gardner

CRC Press is an imprint of Taylor & Francis Group, LLC

Reasonable efforts have been made to publish reliable data and information, but the author and publisher cannot assume responsibility for the validity of all materials or the consequences of their use. The authors and publishers have attempted to trace the copyright holders of all material reproduced in this publication and apologize to copyright holders if permission to publish in this form has not been obtained. If any copyright material has not been acknowledged please write and let us know so we may rectify in any future reprint.

Except as permitted under U.S. Copyright Law, no part of this book may be reprinted, reproduced, transmitted, or utilized in any form by any electronic, mechanical, or other means, now known or hereafter invented, including photocopying, microfilming, and recording, or in any information storage or retrieval system, without written permission from the publishers.

For permission to photocopy or use material electronically from this work, access www.copyright.com or contact the Copyright Clearance Center, Inc. (CCC), 222 Rosewood Drive, Danvers, MA 01923, 978-750-8400. For works that are not available on CCC please contact mpkbookspermissions@tandf.co.uk

Trademark notice: Product or corporate names may be trademarks or registered trademarks and are used only for identification and explanation without intent to infringe.

Library of Congress Cataloging-in-Publication Data (applied for)

ISBN: 978-1-032-27021-0 (hbk)
ISBN: 978-1-032-27003-6 (pbk)
ISBN: 978-1-003-29090-2 (ebk)

DOI: 10.1201/9781003290902

Typeset in Palatino Linotype
by Radiant Productions

Preface

The reason for writing this dissection manual is simple. As anyone in academia will tell you, one always has an itch to contribute to their discipline in some way, either large or small. A scholarly endeavor like this surely scratches that itch.

First, this is a manual for proper cadaver dissection. It is not an atlas of human anatomy, nor is it a treatise of all things human anatomy, and therefore it is to be used in conjunction with any of the excellent anatomy atlases currently available.

Secondly, the purpose of this cadaver dissection manual is to provide a relatively succinct and useful dissection protocol for anatomy students in a professional program. This dissection manual covers almost all regions of cadaver dissection typically done at the graduate and postgraduate level.

It is hoped that this manual combined with its Clinical Applications, will meet the needs of students pursuing degrees in medicine, nursing, physical therapy, dentistry, and any other health profession that requires extensive knowledge of anatomical relationships and how those relationships lead to dysfunction and disability.

I believe that the purpose of teaching anatomy lies in its *clinical application*, thus dissecting cadavers or examining prosected specimens, coupled with **clinical relevance,** is the hallmark of anatomy teaching at the professional level.

Finally, in this manual I have tried to keep in mind that in teaching the anatomical sciences there is a difference between **"Nice to Know"** and **"Need to Know."** I believe clinicians prefer **"Need to Know,"** but academics prefer both. I think it is a very fine line to walk with an endeavor such as this, but I have tried to appease both audiences, the clinicians, and academics. Hopefully, I have succeeded.

Acknowledgements

I am grateful to many of my professors who over the years I had the pleasure of first being a student of theirs, and later working, teaching, and dissecting alongside them in gross anatomy labs. I wish to acknowledge two anatomy professors that have contributed to my anatomical acumen in immeasurable ways, Se-Pyo Hong and Harvey "Chip" Morter both formerly of Palmer College of Chiropractic. I also would like to mention two of my anatomy colleagues at Lake Erie College of Osteopathic Medicine, Dr. Frank Liuzzi and Dr. Oren Rosenthal. I appreciate their friendship and support. Finally, I am grateful to Stephanie Klinesmith of Touro University, College of Osteopathic Medicine for reviewing parts of the manuscript.

Contents

Note: You may not be able to follow Unit 2 dissection sequentially. It will depend upon whether the forearms of the cadaver are supinated or pronated.

How to Use This Manual

The dissection manual is divided into *Units*, beginning with the back and suboccipital region. It is customary (but not required) to begin with the back and move on from there. However, the units can be used on their own too, without following chronological order.

This manual follows an easy-to-use format. Each section includes subheadings and explanations. The subheadings guide you on what to do prior to the dissection, cadaver position, incisions to make and what to look for during the dissection as well as explaining the clinical relevance of anatomy of that region. The last Unit of the manual is designated as Specialty Dissections. They should be done if there is time and interest. They may require unconventional tools such as a bandsaw or scroll saw to complete.

These dissections are very specific and take a lot of patience and skill but are well worth doing!

Explanations of the Subheadings

Preparation

The information found in this section will prepare you for a successful dissection. This section in each unit will explain what you will need to do *prior to the dissection* to make it a success. It is not required nor necessary to spend hours and hours reading the tomes of anatomy prior to dissecting an area. However, prior to every dissection you will need to peruse an anatomy atlas for structure, location, and identification of the regional anatomy. **If you do not know where to look for a structure and what it looks like, you will spend many hours in the lab, all for naught.**

Cadaver Position

This section describes the position the cadaver needs to be placed for a successful dissection. Also, blocking, rope tying, and limb position will be discussed here. Be aware that in some regional dissections you will need to reposition the cadaver from supine to prone and back again.

Pre-dissection Discussion

Included here is a variable discussion of what you can expect to see during the dissections, how long it may take, various anatomical relationships and the many pitfalls to avoid while dissecting each region. It may include lists, memory mnemonics or other useful approaches to the dissections.

Incisions to Make

This portion will describe or show in diagrams the specific cuts you will need to make in the integument, connective tissue, or bone to be able to uncover the anatomical structures of interest. It may also include pictures or drawings indicating the directions to reflect the tissues to clearly present the anatomy of interest.

Structures to Clean and Identify

Included here will be a list of muscles, nerves, blood vessels, connective tissue, and organs in the region to uncover and clean. Spaces, cavities, and membranes will also be noted too. Very brief anatomical descriptions of various structures may be included here as well.

Clinical Applications and Orthopedic Assessment

Here the reader will find a non-exhaustive list of briefly defined clinical terms, clinical disorders, pathognomonic radiological findings, and relevant regional orthopedic tests.

Introduction

A. The Study of Anatomy

Anatomy belongs to the basic science discipline. The science of anatomy concerns the structural organization of the human body. Anatomy is concerned with the study of structure. Most terms and concepts in human anatomy have their origins in Greek or Latin. The term *anatomy* is of Greek origin, meaning *to cut up or dissect.* Anatomy is a descriptive science, meaning most of the terms are understood only if their prefixes and suffixes are defined.

B. There are Many Ways to Study Anatomy

Gross anatomy

The study of anatomy using the unaided eye. This type of anatomical study is most often done in the cadaver lab.

Microscopic anatomy or Histology

The study of minute structures with the aid of a microscope or some other high-powered lens.

Comparative anatomy

The comparative study of structures in animals in relation to each other and to human structures.

Regional anatomy

Studying regions of the body such as the head, neck and trunk. This type of anatomical study is usually done in the cadaver lab during dissection.

Developmental anatomy

The study of the embryonic origin and development of the structures. It is the foundation of gross anatomy. It explains why structures are located in certain positions and from which embryological tissues they are derived.

Functional anatomy

The study of the total interaction of the organ systems in a very dynamic fashion. It can be thought of as the study of the moving parts of the organism.

Pathologic anatomy

The study of the diseased structures, their locations, and subsequent regional effects.

Radiographic anatomy

The study of the human body using X-rays, CT, MR or PET scans, etc. to evaluate and diagnose disease.

There are generally three main approaches to studying anatomy in academia.

1. **Systemic Anatomy** is most often studied in undergraduate anatomy and physiology classes. This type of study allows one to look at the body as a series of systems, including but not limited to the Integumentary, Nervous, Cardiovascular, Endocrine and Skeletal systems.

2. **Regional Anatomy** is best appreciated during cadaver dissection. In this type of study, the cadaver is dissected by region (cervical, abdominal, thoracic, etc.) and all systems and structures are studied at the same time as seen in that region.

3. **Clinical Anatomy** is the act of incorporating both regional and systemic anatomy for the benefit of training clinicians such as doctors, dentists, and other healthcare providers. It is the most useful method of the three, in this author's opinion.

C. Anatomical Regions and Contents

Regions of the Body

Regional dissection is the manner in which anatomy is studied in the cadaver lab, usually. For example, some regions dissected and studied might include:

a. Gluteal region
b. Suboccipital region
c. Popliteal region
d. Axillary region
e. Pectoral region
f. Temporal region
g. Infratemporal region

The abdominopelvic area can be divided into nine regions using four imaginary lines. Two lines will pass from the right and left midclavicular (**midclavicular lines**) areas vertically to the pelvis. Regarding the other pair, one will pass just below the ribcage horizontally at the costal margin (**subcostal line**) while the other will pass horizontally across the hips connecting each iliac tuberosity (**transtubercular line**), thus forming a tic-tac-toe arrangement on the abdominopelvic area. The nine regions formed from these imaginary lines include:

The **umbilical region** centermost region surrounding the umbilicus.

The **epigastric region** located just superior to the umbilical region.

The **hypogastric region** located just inferior to the umbilical region.

The **right and left iliac or inguinal regions** located lateral to the hypogastric regions.

The **right and left lumbar regions** located lateral to the umbilical region.

9 Anatomical Regions and Their Contents

Right Hypochondriac	Epigastric	Left Hypochondriac
Right kidney	Esophagus	Pancreas (tail)
Gallbladder	Stomach	Left kidney
Liver	Pancreas	Descending colon
Ascending colon	Transverse colon	Spleen
Part of duodenum	Liver	Stomach
	Duodenum	Left colic flexure
Right Lumbar	**Umbilical**	**Left Lumbar**
Ascending colon	Ileum	Left kidney
Liver	Jejunum	Descending colon
Gallbladder	Umbilicus	
Right kidney		
Right Iliac	**Hypogastric**	**Left Iliac**
Appendix	Urinary bladder	Sigmoid colon
Cecum	Fundus of uterus	Descending colon
Ileum coils	Ileum	Ovary
Ovary		Oviduct
Oviduct		

Quadrants of the Body

There are four quadrants formed when imaginary lines are passed through the umbilicus. One line is vertical, and the other is horizontal. These two lines, a vertical line running through the middle of the abdomen and a horizontal line through the umbilicus, divide the abdominal-pelvic cavity into four areas or quadrants. These quadrants are described and abbreviated as Right Upper Quadrant (**RUQ**), Right Lower Quadrant (**RLQ**), Left Upper Quadrant (**LUQ**), and Left Lower Quadrant (**LLQ**). These quadrants are most often used clinically to assess and document a patient's abdominal complaint.

4 Abdominal Quadrants and Their Contents

The quadrant system of the abdominopelvic region is most often used clinically, but at times, the nine regions are also used, especially in sonography.

Right Upper Quadrant (RUQ)	Left Upper Quadrant (LUQ)
Right lobe of liver	Left lobe of liver
Gallbladder	Most of stomach
Pyloric part of stomach	Jejunum
Head of pancreas	Body and tail of pancreas
Part of transverse colon	Part of transverse colon
Right colic flexure	Left colic flexure
Right kidney	Left kidney
Most of duodenum	Left suprarenal gland
Right suprarenal gland	
Right Lower Quadrant (RLQ)	**Left Lower Quadrant (LLQ)**
Most of ileum	Sigmoid colon
Part of ascending colon	Part of descending colon
Cecum and vermiform appendix	Ovary and oviduct
Ovary and oviduct	Left ureter
Right ureter	

D. Important Planes of the Body

There are four (4) *useful* anatomical planes of the body created by imaginary lines passing through the body. They include:

1. **Sagittal Plane** (Para-sagittal Plane) a plane that divides the body into right and left areas, not usually equally, unless it is the mid-sagittal plane. *See below*

2. **Mid-sagittal Plane** (Median) a plane that divides the body into *equal* right and left halves.

3. **Frontal or Coronal plane** a plane that divides the body into anterior and posterior parts, not necessarily equally.

4. **Transverse plane** (aka Cross "X" Section) a plane that divides the body into superior and inferior portions, again, not usually equally.

Anatomical Terminology

Using the proper terminology is the key to clear communication in anatomy. Anatomists have their own lexicon. In order to communicate with anyone in the field of anatomy or anyone in healthcare, one must learn the language. It is imperative to know the meanings and proper uses of the anatomical terms, or one will surely be lost. In this manual I have striven to use anatomically correct terminology and not colloquial terms.

I have on occasion used various eponyms, however.

Anatomical Position

Even though the cadaver is lying on a table, the anatomical position still applies and governs *all* directional terminology and anatomical relationships.

The anatomical position is the starting point for all things anatomy. The anatomical position is described below.

A person is erect and standing with their feet shoulder width apart, arms at the sides with palms facing forward. The head is positioned in the neutral position with eyes looking forward.

Directional Terms

Term	Definition	Example
Superior	Toward the head	The diaphragm is superior to the liver
Inferior	Away from the head	The liver is inferior to the diaphragm
Anterior	Toward the front	The sternum is anterior to the heart
Posterior	Away from the front	The heart is posterior to the sternum
Medial	Toward the midline	The nose is medial to the ears
Lateral	Away from the midline	The ears are lateral to the nose
Ipsilateral	On the same side	The gallbladder is ipsilateral to the liver
Contralateral	On the opposite side	The spleen is contralateral to the gallbladder
Proximal	Closer to the attachment of a limb to the body	The elbow is proximal to the wrist
Distal	Further away from the attachment a limb to the body	The wrist is distal to the elbow
Superficial	Closer to the surface of the body	The skin is superficial to the muscles
Deep	Further away from the surface of the body	The muscles are deep to the skin

Osseous Surface Markings

All bones have at least one bony landmark. Most will have many. It is extremely important to know when to use each term and to what each term refers. Some terms are used interchangeably, most are not. Generally speaking, the bony landmarks provide for muscle and ligament attachments as well as serving as passageways for nerves, blood vessels and lymphatics.

Term	Definition	Example
Fossa	A shallow depression	Glenoid fossa
Fovea	A pit	Fovea capitis
Foramen	A hole	Foramen magnum
Foramina	Plural of foramen	Olfactory foramina
Fissure	Narrow slit	Superior orbital fissure
Sulcus	Shallow groove	Bicipital sulcus
Crest	Prominent ridge	Iliac crest
Condyle	Large, rounded protuberance	Occipital condyle
Epicondyle	Rounded projection above a condyle	Medial epicondyle
Process	General term for a projection	Spinous process
Trochanter	A very large projection	Greater trochanter
Tubercle	Small, rounded projection	Lesser tubercle of humerus
Tuberosity	Small, roughened projection	Lesser tuberosity of humerus
Facet	Smooth articular surface	Superior articular facet
Head	A rounded projection on the epiphysis	Humeral head
Neck	Narrow, constricted portion of bone	Neck of radius
Suture	Seam-like articulation	Coronal suture
Meatus	Tube-like opening	Internal acoustic meatus
Linea	A line, often roughened	Linea aspera
Protuberance	A rounded projection	External occipital protuberance
Canal	A narrow, enclosed pathway	Infraorbital canal
Groove	An uncovered narrow slit	Infraorbital groove

Muscle Terminology and Examples

Muscles are named by a variety of methods, including their actions, their size, their fascicle direction, number of bellies they possess, their location as well as by their attachments to bony landmarks. The terms used often reflect these arrangements.

Term	Definition	Example
Rectus	Straight fiber orientation or parallel to midline	Rectus abdominis
Transverse	Fibers are perpendicular to midline	Transverse abdominis
Oblique	Fibers are diagonal to midline	External abdominal oblique
Maximus	Largest in size	Gluteus maximus
Medius	Intermediate in size	Gluteus medius
Minimus	Smallest in size	Gluteus minimus
Longus	Long	Adductor longus
Brevis	Short	Adductor brevis
Latissimus	Wide	Latissimus dorsi
Longissimus	Longest	Longissimus capitis
Magnus	Largest	Adductor magnus
Major	Larger	Pectoralis major
Minor	Smaller	Pectoralis minor
Vastus	Huge	Vastus lateralis
Deltoid	Triangular	Deltoid
Trapezius	Trapezoid-like	Trapezius
Serratus	Saw-toothed	Serratus anterior
Rhomboid	Rhombus-like	Rhomboid major
Orbicular	Circular in shape	Orbicularis Oris
Pectinate	Teeth of a comb-shaped	Pectineus
Piriform	Pear-shaped	Piriformis
Playts	Flat	Platysma
Quadratus	Four-sided	Pronator Quadratus
Gracile	Slender	Gracilis
Flexion	Decrease the angle at a joint	Flexor carpi radialis
Extension	Increase the angle at a joint	Extensor carpi radialis
Abduction	Move away from the midline	Abductor digiti minimi
Adduction	Move toward the midline	Adductor Pollicis

Table contd. ...

...Table contd.

Term	Definition	Example
Levator	Move a body part superiorly	Levator scapulae
Supination	Turn the palmar surface anteriorly	Supinator
Pronation	Turn the palmar surface posteriorly	Pronator teres
Rotator	A muscle that moves a bone around its longitudinal axis	Rotator longus
Biceps	Refers to a muscle with two origins	Biceps brachii
Triceps	Refers to a muscle with three origins	Triceps brachii
Quadriceps	Refers to a muscle with four origins or gasters	Quadriceps femoris
Sphincter	A muscle that regulates the size of an opening	External anal sphincter

E. Before You Begin

Always Prepare for the Dissection!

It is best to study the area to be dissected *prior* to beginning the dissection by reading the appropriate part of this manual *AND* using a good human anatomy atlas for reference. If you have not prepared for the dissection, you can be assured of a catastrophic result!

Dissecting a cadaver takes time and a certain amount of honed skill. It will take much practice and patience to become competent in cadaver dissection. Before beginning any dissection, it is important to know the answers to these three questions:

1. **What are you looking for?**
2. **Where should it be located?**
3. **What does it look like?**

For instance, if one is looking for the radial nerve as it runs with the brachial profundus artery in the posterior arm, it is important to look within the Triangular Interval. The Interval is an opening between the long and lateral heads of the triceps brachii muscle. The radial nerve is usually large, flat, and quite shiny, while the brachial profundus artery is typically smaller, duller, cylindrical and cord-like.

Again, if you do not know what you are looking for, where it should be located and what it looks like, you will spend many, many hours in the lab all for naught. Be prepared!

F. Care of the Cadaver

It is imperative that the cadaver not dry out! The cadaver will require constant monitoring to prevent drying out. First, only uncover the area on the cadaver that you are working on during that day's dissection. Then, once you are done dissecting for the day, be sure to moisten the cadaver with the wetting solution provided and cover the cadaver. Doing this routine proper care will provide for many, many months of rewarding dissection. **Believe it or not, if properly cared for, some cadavers can last for more than 20 years!**

It is customary to keep the hands, feet and face covered until you are ready to dissect those areas. It is a courtesy to the cadaver as well as to your classmates who may feel uncomfortable with those areas exposed. **At no time should the face of the cadaver lie directly on the dissection table. It is disrespectful.**

Most of the time the skin will be kept attached to the cadaver to be used as cover to prevent drying out. Later in the dissection, the skin may be discarded to clean up the dissection field and make it easier to manipulate the cadaver for presentation and study.

Tip: This author has found that twin-sized bed sheets work very well in covering the cadaver. First, they are moistened with the wetting solution, draped over the cadaver and then the cadaver is wrapped tightly in plastic.

Proper Dissection Tools and Protective Gear

Lab coat
Eye protection
Gloves
Blunt probe
Scalpel (and a good supply of disposable blades)
Small and large scissors
Hemostats of varying sizes
Autopsy saws
A small hand-held rotary cutting tool

Various sized chisels and mallets

A dissection light or small flashlight

A length of rope

Various sized storage containers with tight lids

A grease pencil for drawing on the skin

G. Dissection Techniques, Tips and Terms

Dissection Tips

Dissection Tip 1

Dissection in a professional program is a *team effort*. Therefore, it is imperative that you choose your partners very carefully. Select a partner or team that is laser-focused on learning as much as they can in the time that is allotted for that day's dissection. If you do not choose wisely, you will forever be playing "catch-up."

Dissection Tip 2

Do not cut too deeply! It is imperative that the skin and superficial fascia be removed in two steps. **First skin, then superficial fascia. Always!**

Dissection Tip 3

Creating a "button-hole" in the skin to insert your finger can allow you to better grasp the skin and traction it while skinning. By the way, once the skin is removed from the cadaver, the blunt probe, hemostats and *your own hands* become the most useful tools for dissection.

Dissection Tip 4

One of the best tips in using the hemostats, is to place the tips of the hemostats in-between the structures you wish to separate and open it slowly. It is called the *"scissors technique"* but using a hemostat. This technique will provide for gentle separation of structures with minimal damage to the surrounding structures.

Dissection Terms

Reflect to move a structure out of the dissection field either by cutting it or simply moving it without cutting it in order to not obscure other structures.

Blunt dissection using either your fingers or blunt probe to clean the dissection field. A blunt probe is used instead of a sharp probe so as to not accidentally destroy important structures.

Preserve to save, but not remove or cut.

Neurovascular bundle implies nerve, artery, veins and lymphatics.

Trajectory the pathway that a structure is taking.

Clean the structure or the dissection field to remove fascia, fat, or some other tissue that is obscuring the field of view.

Medial and lateral attachments These terms are used instead of origin and insertion. They are not interchangeable with origin and insertion but used here to de-emphasize the importance of memorizing each muscle's origin and insertion. Medial is often thought of as being the origin, while lateral is often thought of as being the insertion of a muscle.

Remember, once the skin is removed, your best tools for dissection will be your own hands! A dissection field that is clean with clearly defined anatomical structures is a sight to behold, one you will surely remember and appreciate.

On the other hand, a poorly defined field with undefined anatomical structures is not a sight to behold and one that you will most certainly want to forget. In fact, it is disrespectful to the cadaver and wasteful if one does not dissect properly.

Finally, this cadaver is your first patient. Treat them with respect and dignity at all times.

Because of their donation, they have allowed you to become a better healthcare provider.

Always be thankful for their donation to science.

UNIT 1
The Back

A. Superficial and Deep Back

1. Preparation

Using a good atlas, begin your preparatory work by studying the relevant anatomy of the vertebral column, posterior skull, posterior neck, and back muscles (superficial, intermediate, and deep), as well as the nerves, ligaments and blood vessels of the area. Do not forget to include the scapula and pelvis in your preparation.

2. Cadaver Position

Place the cadaver in the prone position (face down) with a block under the sternum keeping the face off of the dissection table. The neck should be in mild anterior flexion.

3. Pre-dissection Discussion

The skin is very thick in the upper back, so you will need to apply some pressure to the scalpel as you open this region. After removing the skin and superficial fascia from the back, you will notice superficial dorsal cutaneous nerves; we will ignore these as they have little clinical importance in dissection.

Next you will see the **trapezius muscle**. Clean the trapezius and reflect it superolaterally. On the deep surface of the trapezius, notice an area of adipose tissue, this fat pad will house the **spinal accessory nerve**, as well as the **superficial branch of the transverse cervical artery**. You also will see the **rhomboid major** and **minor** deep to this muscle as well as the **Splenius capitis** and **Splenius Cervicis**, a "bandage-like" muscle. Do not dissect these yet.

Next, reflect the **Latissimus Dorsi** muscle. It is a large, flat, fairly thin muscle, running from T7 to the sacrum via the Thoracolumbar fascia. Clean the muscle and reflect it laterally. You must use your hands to separate it from the underlying ("paper-thin") **Serratus Posterior Inferior** muscle. Locate branches of the **Thoracodorsal Nerve and Artery** on the deep surface of latissimus dorsi muscle and preserve them. This neurovascular bundle will be discussed in a later dissection. This neurovascular bundle can be found in the pocket between the latissimus dorsi muscle and the serratus anterior muscle near the inferior angle of the scapula.

Clean and reflect the Rhomboids and cut them at their medial attachments to the vertebrae and appreciate the "paper-thin" **Serratus Posterior Superior** muscle just deep to them. Cut this thin muscle at its medial attachments and reflect it laterally.

The **Dorsal Scapular Nerve and Artery** travels parallel to the medial border of the scapula innervating and supplying blood to the Rhomboids and the **Levator Scapulae.** Cut the levator scapulae muscle at its scapular attachment and reflect it superiorly.

Examine the posterior neck. Notice the Splenius (a bandage-like) muscle. It is composed of the Splenius Capitis (the large superolateral portion) and the Splenius Cervicis (small inferolateral portion) muscles. These are the two parts of this muscle that lie deep to the trapezius. You may also notice at this time, the Longissimus Cervicis just lateral to the Splenius Cervicis, ignore it for now. Clean both the **Splenius Capitis** and **Splenius Cervicis** and reflect them. Beneath this muscle you will see the semispinalis capitis muscle with the **Greater Occipital Nerve, Occipital Artery** and **Third Occipital nerve** piercing it. Do not confuse the occipital artery with the **Posterior Auricular Artery**. The posterior auricular artery is much more laterally placed than the occipital artery and runs directly behind the ear. It is much smaller than the occipital artery.

The **Erector Spinae** group of muscles are composed of the Iliocostalis (lateral group), Longissimus (medial group) and the Spinalis (median group). Each of these may possess a regional group named appropriately, such as Iliocostalis cervicis and longissimus capitis, etc. *These subgroups will not be dissected here,* but they should be studied in an atlas. These muscles and their function can be remembered using the mnemonic "I Like Standing" (Iliocostalis, Longissimus, Spinalis keep you up-right).

Lastly, the deepest group is composed of the Semispinalis group, Multifidi, Interspinous, Intertransversarii, Rotatores, and Levator Costarum (ribs) muscles (brevis and longus). They all lie deep to the Erector Spinae and fill in the spaces between the spinous processes and transverse processes of the vertebrae, except the levator costarum, attached to the ribs. **These are not to be dissected here.** Find those in an atlas.

4. Incisions to Make

Make a superficial cut from the external occipital protuberance (EOP) to the sacrum, then make several additional cuts perpendicular from this cut to about the mid-axillary line then continue them out to the tip of the shoulder, and then cut from the mastoid process to the EOP. Also, make a cut from the lateral sacrum up and over the outline of the gluteus maximus. Reflect the skin flaps from medial to lateral. **Do not** remove the skin from the cadaver completely. Leave the skin flaps attached laterally to use as cover to prevent drying out. **See Figure 1.**

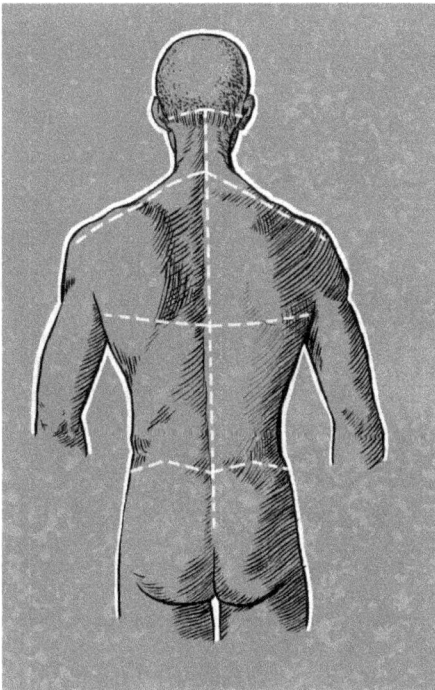

Figure 1. Incisions to make on the back.

5. *Structures to Clean and Identify* (See Figure 2, Figure 3 and Figure 4)

Muscles

Trapezius the most superficial muscle of the back and presents superior, middle and inferior fibers. Identify these fibers by looking at fiber direction.

Latissimus Dorsi is the "swimmer's muscle," the wide muscle of the back giving the typical "V" shape in trained individuals.

Rhomboid major & minor these two muscles are not always clearly separated in the cadaver

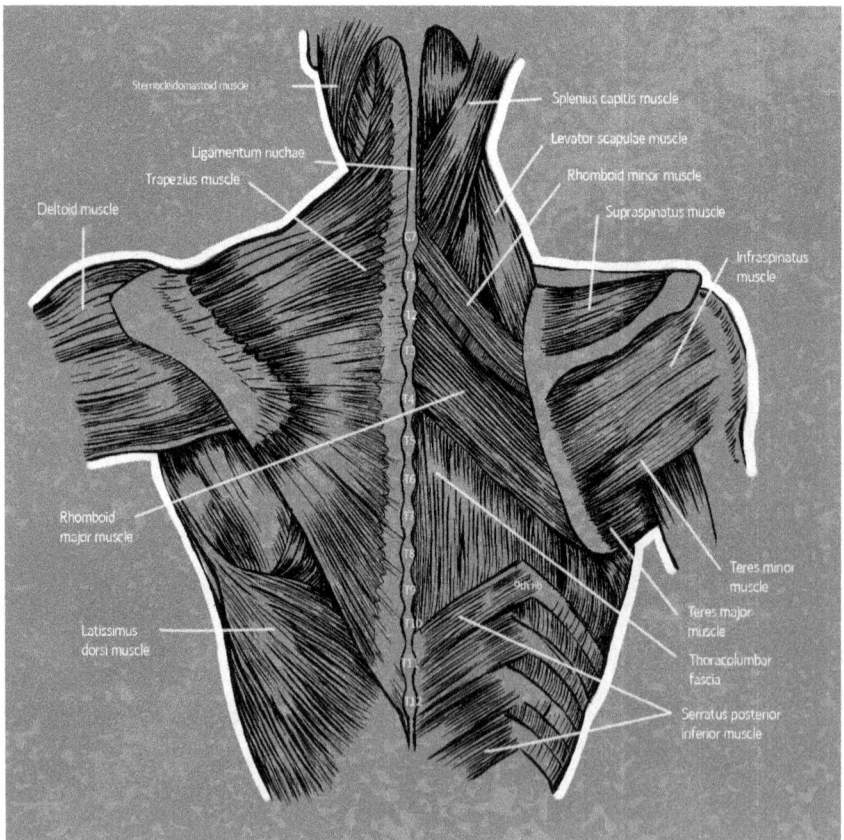

Figure 2. Superficial dissection of upper back.

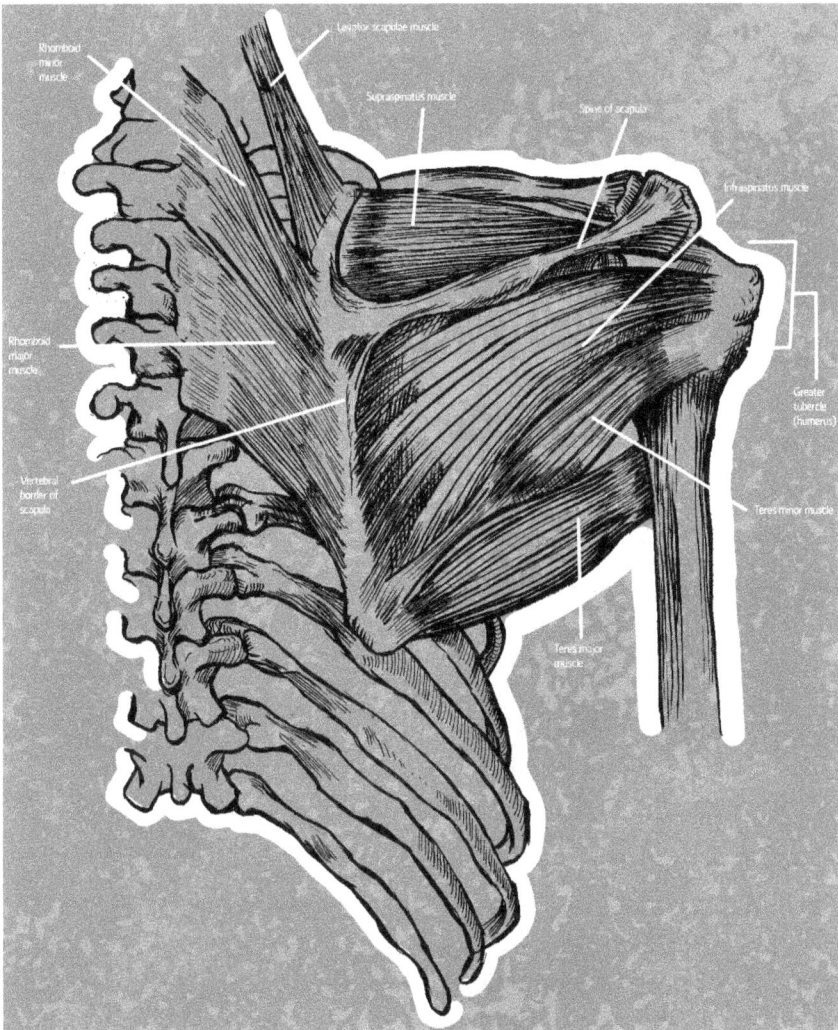

Figure 3. Deep dissection upper back. The trapezius muscle has been removed.

Levator scapulae notice four slips of origin of this muscle from the upper cervical vertebrae. This muscle is highly variable. You may notice more or less attachments.

Serratus posterior superior & inferior "paper-thin" muscles that often adhere to the muscles superficial to them.

Thoracolumbar fascia a very thick and dense layer of "glistening" connective tissue attaching medially to the spines of the lumbosacral area and providing attachment of the latissimus dorsi muscle.

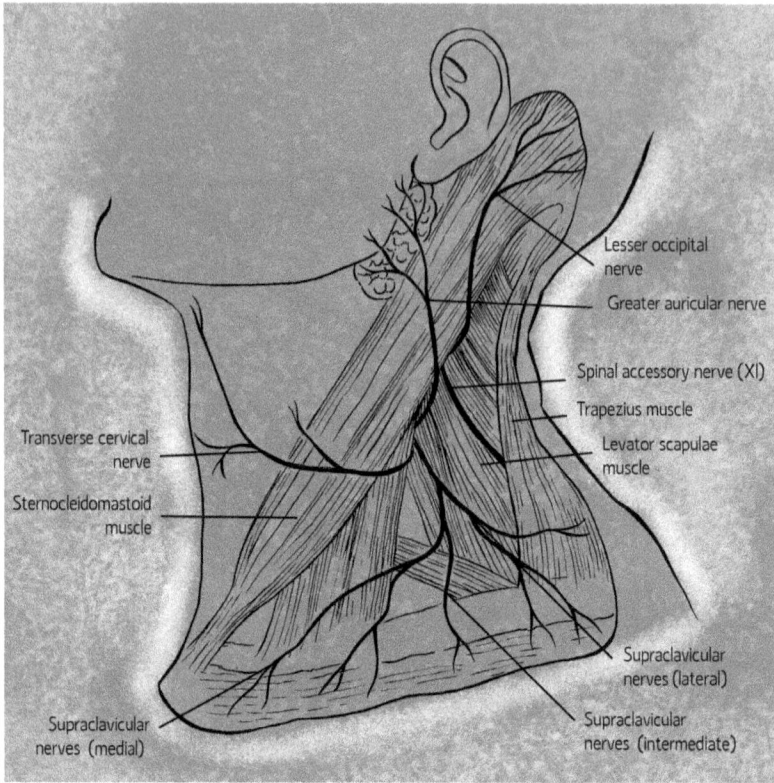

Figure 4. Superficial dissection of lateral neck showing nerves of cervical plexus.

Splenius capitis this muscle attaches to the mastoid process, hence the name.

Splenius cervicis these muscles are actually two parts of the same muscle, and they will be reflected together in the dissection. It does not attach to the mastoid process.

Semispinalis capitis the largest muscle mass of the deep posterior neck. It is a key muscle to appreciate when dissecting the suboccipital triangle. The greater occipital nerve pierces it.

Nerves

Dorsal cutaneous branches of segmental spinal nerves

Spinal accessory the eleventh cranial nerve. It innervates the sternocleidomastoid and the trapezius muscles.

Greater occipital (C2 dorsal ramus) it is purely sensory to the back of the head, almost up to the vertex.

Lesser occipital (C2 ventral ramus) located along posterior border of the sternocleidomastoid (SCM)

Least or 3rd occipital (C3 dorsal ramus) located just inferior to the C2 spinous process running medially to the greater occipital nerve initially.

Dorsal scapular (C5 root) often seen just medial to the levator scapulae running along the medial border of the scapula.

Thoracodorsal (C6–C8) also known as the middle subscapular nerve. It is from the posterior cord of the brachial plexus and innervates the latissimus dorsi muscle.

Blood vessels

Superficial and Deep branches of the Transverse cervical artery from the Thyrocervical trunk of the Subclavian artery.

Note: The superficial branch is seen on the posterior surface of the trapezius while the deep branch is sometimes known as the dorsal scapular artery and is seen running with the dorsal scapular nerve. To accurately differentiate these vessels, they would need to be traced back to their branching point from the subclavian artery. We will see this in detail later.

Dorsal scapular artery directly from the subclavian artery (usually).

Occipital artery from the external carotid artery and is positioned *lateral* to the greater occipital nerve passing through the substance of the semispinalis capitis muscle on its way to the vertex of the skull.

Posterior Auricular artery a small branch from the external carotid artery that travels directly behind the ear.

6. *Clinical Applications*

a. **Triangle of Auscultation** formed by the superior border of latissimus dorsi, lateral border of trapezius and medial border of scapula (the rhomboid major forms the floor). This triangle allows very clear auscultation of breath sounds.

b. **Lumbar "Petit's" Triangle** formed by the borders of the latissimus dorsi, external abdominal oblique muscle and illiac crest. The floor is formed by the internal abdominal oblique muscle. The lumbar triangle can be a location of a hernia of retroperitoneal fat, kidney, or colon.

c. **Intervertebral Disk Herniation** a condition affecting the spine in which the annulus fibrosus is damaged enabling the nucleus pulposus to herniate often posterolaterally, compressing nerves or the spinal cord causing pain and dysfunction.

d. **Military neck** loss of the normal lordosis of the cervical spine from degeneration or trauma

e. **Scoliosis** an abnormal lateral curvature of the spine quantified using Cobb angle. It is mostly seen in adolescent girls and is idiopathic.

f. **Spondylosis** a degeneration of the spinal column usually presenting with osteophytes

g. **Spondylolyis** a defect in the pars interarticularis of a vertebral arch

h. **Spondylolisthesis** anterior slippage of one vertebra relative to the one below it. Most commonly seen at L5-S1. Severity is graded using Meyerding scale

i. **Spinal stenosis** a narrowing of the vertebral canal often due to thickening of the ligamentum flavum

j. **Spina bifida** an incomplete fusion of the posterior element of a vertebra of varying severity. It can go undetected for years if it is spina bifida occulta. Commonly presents with increased hair growth over the affected area.

k. **Dowager's hump** a hyperkyphosis of the thoracic spine often due to multiple compression fractures. Especially noted in elderly women.

l. **Winking Owl sign** absence of a pedicle on plain radiograph. Strongly suggestive of cancer or infection.

m. **Collared scotty dog** fracture of the pars interarticularis seen on plain X-ray film

n. **Clay shoveler's fracture** a fracture of the spinous process of lower cervical vertebrae from trauma. It is generally a stable fracture that is unlikely to need surgical treatment.

B. SubOccipital Region

1. *Preparation*

Begin this dissection by studying the posterior skull, the cervical vertebrae, the origin and trajectory of the vertebral artery as well as the cervical plexus in an atlas. You will need a small flashlight or some other light source for this dissection.

2. *Cadaver Position*

The cadaver should be prone (face down) with a block under the sternum to keep the face off of the dissection table. The head should be in mild anterior flexion.

3. *Pre-dissection Discussion*

The Suboccipital Region comprises of the Suboccipital Triangle, the deepest area of the posterior neck. These deep muscles connect the atlas (C1) and axis (C2) to each other as well as to the posterior skull. Two muscles in this region are designated as "oblique" (obliquus) and two muscles are designated as "straight" (rectus) relative to their fascicle orientation to the long axis of the body. Once the Semispinalis Capitis muscle is cut and reflected superiorly and/or inferiorly you will notice the following suboccipital muscles:

Rectus Capitis PosteriorMajor (RCPMj)
Rectus Capitis PosteriorMinor (RCPMi)
Obliquus Capitis Superior (OCs)
Obliquus Capitis Inferior (OCi)

The *Suboccipital Triangle* is composed of 3 of the 4 above noted muscles:
OCs, OCi and RCPMj muscles.

When you puncture the very tough and strong Posterior Atlanto-Occipital Membrane, which forms the floor of the Triangle, you will see the Vertebral Artery as it courses up and around the posterior arch of atlas in the sulcus for it on the posterior arch. You will also see the SubOccipital Nerve within the triangle, often medial to the OCs. Be very careful to preserve the Occipital artery and the Greater Occipital nerve. The greater occipital nerve will be large and appear to be passing superiorly over the belly of the OCi muscle.

4. Incisions to Make

Section the Semispinalis about 2" inferior to the EOP and reflect it superiorly and inferiorly. Alternatively, you may also remove its attachment to the EOP and simply reflect it inferiorly. Use the blunt probe, the scissors technique with the hemostats, and/or your fingers to clean the dissection field. However, the deep fascia is very thick in this region, so snipping with scissors *may be necessary* to clear the dissection field in order to clearly see the muscles.

5. Structures to Clean and Identify (See Figure 5)

Muscles

Rectus Capitis Posterior Major (RCPMj)
Rectus Capitis Posterior Minor (RCPMi)
Obliquus Capitis Superior (OCs)
Obliquus Capitis Inferior (OCi)

This muscle does not attach to the skull, and it may need to be reflected to see the vertebral artery.

Connective tissue

Posterior Atlanto-Occipital Membrane

It forms the *floor* of the suboccipital triangle. The vertebral artery will pierce this membrane. It connects the posterior portion of the foramen magnum to the posterior arch of C1, in particular, the arcuate rim. The *"arcuate rim"* describes the superior margin of the posterior arch of C1, exclusive of the groove for the vertebral artery.

Nerves

Suboccipital Nerve (C1 dorsal ramus) it does **not** have a dermatome

Greater Occipital Nerve (C2 dorsal ramus) it is purely sensory to the back of the head, almost up to the vertex. it is large and can be seen around the lower border of the OCi as it crosses over the suboccipital triangle.

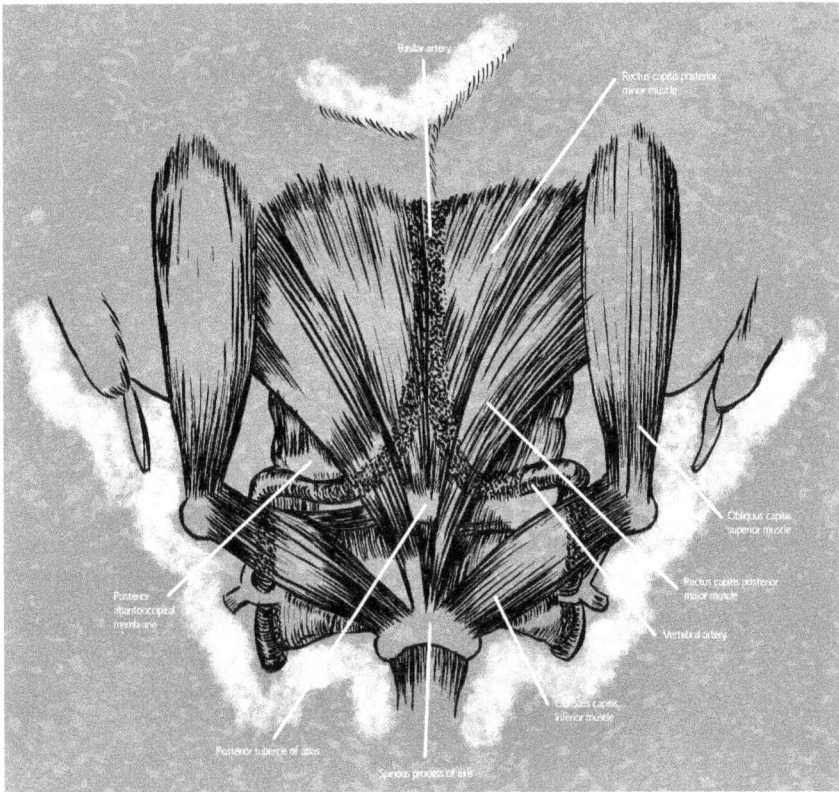

Figure 5. Dissection of suboccipital triangle.

Blood Vessels

Vertebral artery typically, the first branch of the subclavian artery that traverses the transverse foramina of C6 and above. It eventually enters the foramen magnum to fuse with its mate to form the basilar artery on the ventral aspect of the pons.

Occipital artery The largest posterior branch from the external carotid artery. It will be found lateral to the Greater occipital nerve.

Posterior Auricular artery a very small branch from the external carotid artery which courses directly behind the ear.

6. Clinical Applications

a. **Atlanto-Dental Interspace** *(ADI)* it is the distance between the fovea dentis on the atlas and the facet for fovea dentis on the odontoid process, usually noted on a plain X-ray film. Greater distances are seen in Down Syndrome and Rheumatoid Arthritis.

b. **Occipital Neuralgia** sudden severe pain radiating up from the occipital region to the posterior scalp originating from the occipital nerves.

c. **Ponticulus Posticus aka "Ponticle" and the Arcuate Foramen** ossification of the posterior atlantooccipital membrane seen on plain film running from the posterior arch of atlas to the foramen magnum creating an opening, the arcuate foramen, a passageway for the vertebral artery.

d. **Os Odontoideum** a small well corticated ossicle at the superior aspect of a hypoplastic dens. It may or may not be pathological.

e. **Persistent ossiculum terminale** failure of fusion of a secondary ossification center of the dens. It is not pathological.

f. **Hangman's Fracture** bilateral lamina and pedicle fracture of the axis vertebra.

UNIT 2
The Upper Limbs

A. Scapular Region

1. Preparation

Begin your preparatory work by reviewing in an atlas the bones of the upper limb including the scapula, clavicle, wrist, and hand. Also, review the muscles of the upper limb, including shoulder girdle muscles and related blood vessels. Be sure to familiarize yourself with the nerves of the brachial plexus, both its terminal and collateral branches.

2. Cadaver Position

Place your cadaver in the prone position (face down) with a block under the thorax to keep the face off of the dissection table. You may have to abduct the arm several times to loosen the glenohumeral joint and the scapulothoracic joints prior to making any incisions.

3. Pre-dissection Discussion

You will remove the fascia from the deltoid and detach the muscle from the scapular spine, but keep it attached anteriorly on the clavicle. When you delicately reflect the deltoid from its scapular attachment, you should uncover the quadrangular space. Note the axillary nerve and its muscular branches as well as the posterior humeral circumflex artery. Just inferomedial to the quadrangular space; you will also uncover the **Triangular Space**. Note the **Circumflex Scapular Artery** within it.

During this dissection you will see three of the four rotator cuff muscles, including **supraspinatus, infraspinatus,** and the **teres minor**. You will not see the 4th yet, the subscapularis, as it is deep to the serratus anterior muscle on the ventral aspect of the scapula within the subscapular fossa.

You will be cutting and removing a small part of the supraspinatus muscle as it lies in the supraspinous fossa. In doing so, you will uncover the **Superior Transverse Scapular Ligament**. You should notice the **Suprascapular A**rtery (going over) and the Suprascapular **N**erve (going under) the ligament.

A helpful mnemonic is *"Army goes over the ligament, Navy goes under the ligament."*

4. Incisions to Make

Some incisions may have already been made in previous dissections. Extend your earlier incision from the acromion process midline all the way to the carpus of the posterior aspect of the forearm. Remove the skin both laterally and medially exposing the entire posterior arm and forearm. You may completely remove the skin here to better expose the structures.

5. Structures to Clean and Identify in the Scapular Region (See Figure 6 and Figure 7 and Figure 8)

Muscles

Deltoid has an anterior, intermediate, and posterior fibers each lending to its diverse action on the humerus

Supraspinatus the most commonly injured rotator cuff muscle. It abducts the arm up to 15 degrees.

Infraspinatus a very powerful lateral rotator of the arm

Teres Minor it is in the same fascia as the infraspinatus and can sometimes be difficult to fully appreciate.

Teres Major helps to form the floor of the axilla

Long, Medial, and Lateral heads of the Triceps Brachii Primary extensor at the elbow joint. Forms two of the three borders of the triangular interval.

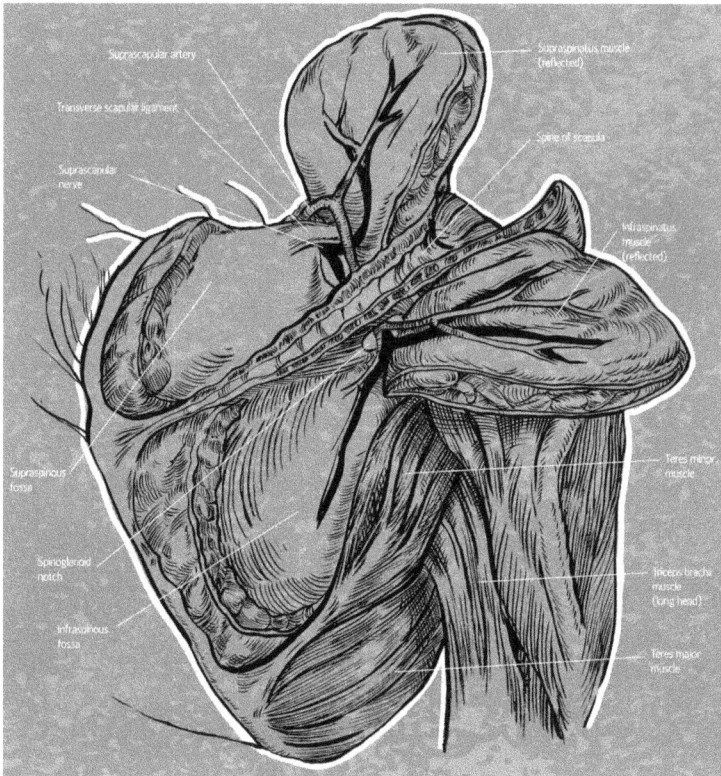

Figure 6. Deep dissection showing quadrangular space. Supraspinatus muscle is reflected.

Connective tissue

Superior Transverse Scapular Ligament this ligament is very strong and extends across the scapular notch on the bony scapula to form a foramen and can be a source of suprascapular nerve compression.

Inferior (spinoglenoid) Transverse Scapular Ligament this ligament overlies the spinoglenoid notch and can also be a source of suprascapular nerve compression

Spaces

Quadrangular Space a 4-sided space bound by teres minor muscle superiorly, the superior border of the teres major muscle inferiorly, the surgical neck of the humerus laterally, and by the long head

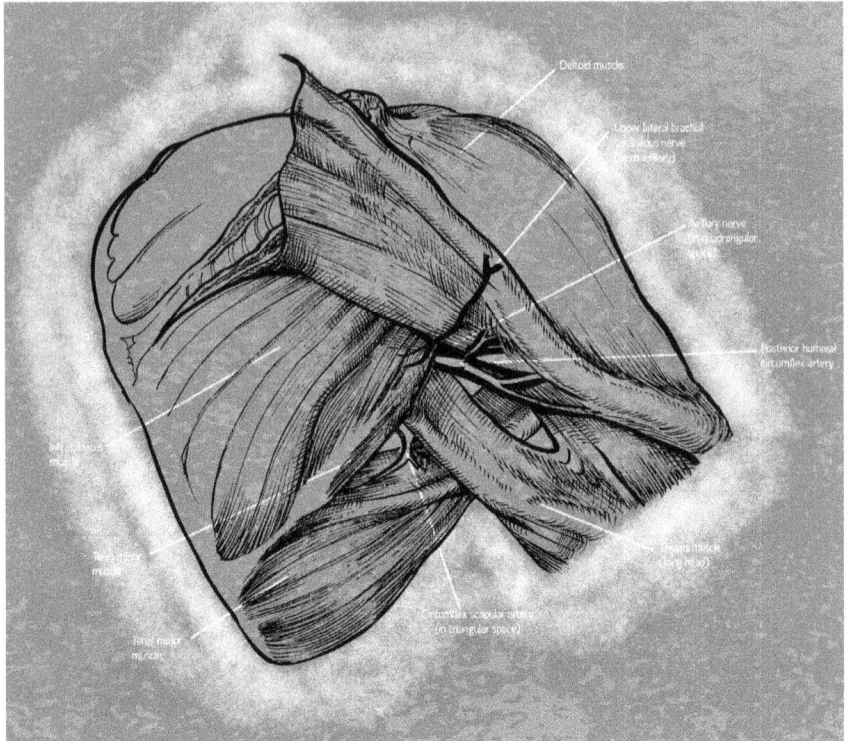

Figure 7. Triangular space.

of the triceps brachii medially. It contains the posterior humeral circumflex artery and the axillary nerve. Note the proximity of the axillary nerve to the second muscle it innervates: the teres minor muscle.

Triangular Space a 3-sided space bound by the teres minor muscle superiorly, teres major muscle inferiorly and long head of triceps brachii laterally. The circumflex scapular artery can be seen here, which is a branch of the subscapular artery.

Triangular Interval a 3-sided interval bound laterally by the lateral head of the triceps, medially by the long head of the triceps and superiorly by the teres major. It contains the brachial profundus artery and the radial nerve.

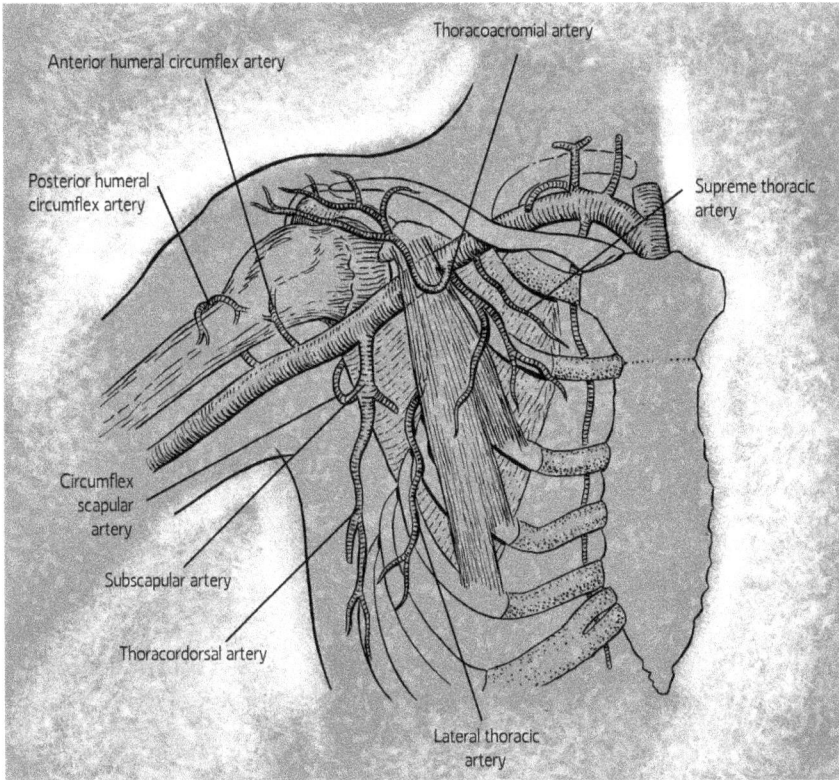

Figure 8. The axilla opened anteriorly. Pectoralis major has been reflected.

Nerves

Axillary nerve (C5–C6) from the posterior cord of the brachial plexus.

Suprascapular nerve (C5–C6) from the upper trunk of the brachial plexus.

Blood Vessels

Suprascapular Artery a branch from the Thyrocervical trunk of Subclavian Artery.

Posterior Humeral Circumflex Artery a branch from the Axillary artery.

Circumflex Scapular Artery a branch of the Subscapular artery from the Axillary artery.

6. Clinical Applications

a. **Quadrangular Space Syndrome** compression of the neurovascular bundle of axillary nerve and posterior humeral circumflex artery often due to trauma or hypertrophy of surrounding muscles

b. **Impingement Syndrome** generally, the entrapment of musculoskeletal structures in the shoulder. It may often affect the tendon of the supraspinatus muscle.

B. Posterior Arm, Posterior Forearm, Dorsum of Hand

Note: Depending on the position of the upper extremity of the cadaver, whether it is supinated or pronated, you may have to skip to the ventral forearm and hand portion of the manual, and *then return* to this point afterwards.

1. Preparation

Begin your preparatory work by studying the bones and landmarks of the upper limb including the bones of the wrist and hand. Also, review the muscles of the upper limb, the nerves of the brachial plexus, as well as the blood vessels of the area.

2. Cadaver Position

Place your cadaver in the prone position (face down) with a block under the thorax to keep the face off of the dissection table. You may have to abduct and medially rotate the arm several times to loosen the glenohumeral joint and upper extremity prior to making additional incisions.

3. Pre-dissection Discussion

After removing the deep fascia, you will see the three heads of the triceps brachii muscle. Separate the long and lateral heads of the triceps muscle. Here you have exposed the triangular interval. Note the radial nerve and the profunda brachii traveling distally within the radial, or spiral groove, of the humerus. Also, be careful of the Ulnar Nerve, which courses very superficial on the distal posterior arm around the medial epicondyle. Extensors of the wrist and fingers and the supinator, in addition to the deep branch of the radial nerve will be identified. Several superficial nerves and veins will also be uncovered too. The extensor retinaculum must be severed on one extremity to study the deep group of muscles.

4. Incisions to Make

Make a longitudinal incision from the acromion to the carpus removing the skin, then the subcutaneous tissue, or extend your earlier incision from the acromion process midline all the way to the carpus of the posterior aspect of the forearm. Sever the extensor retinaculum in the midline and retract it. Remove the skin exposing the entire posterior arm and forearm, you may remove it completely, but do your best to keep it in large pieces for use as a covering when done.

5. Structures to Clean and Identify

Muscles

Triceps brachii (its three heads)

Anconeus a small insignificant triangular muscle on the posterolateral aspect of the elbow that assists the triceps in forearm extension.

Extensor retinaculum a thickened sheet of connective tissue that prevents bowstringing of the extensor tendons.

Common Extensor Tendon attaches the extensor muscles to the lateral epicondyle.

Brachioradialis covers the superficial branch of the radial nerve in the forearm. Commonly used to differentiate flexor and extensor compartments of the antebrachium.

Extensor Carpi Radialis Longus and Brevis
Extensor Digitorum
Extensor Indicis
Extensor Digiti Minimi
Extensor Carpi Ulnaris

Supinator the deep branch of the radial nerve pierces this muscle and when it emerges it is then known as the Posterior Interosseous Nerve.

Abductor Pollicis Brevis the most superficial muscle of the thenar eminence.

Extensor Pollicis Longus and Brevis

Anatomical Snuff Box bound *medially* by the tendon of the extensor pollicis longus and *laterally* by the tendons of the extensor pollicis brevis and abductor pollicis longus, creating a "brevis sandwich" The scaphoid bone forms the major portion of the floor of the Anatomical Snuffbox along with the Trapezium. The radial artery and superficial branch of the radial nerve traverse this box superficially.

Nerves

Radial Nerve (C5-T1)

a. **Deep branch of radial nerve** referred to the "posterior interosseous nerve" *after* it passes through the supinator muscle.

b. **Superficial branch of radial nerve** it emerges at the distal part of the forearm undercover of the brachioradialis and crosses the Anatomical Snuffbox. It is a fairly large structure as it approaches the lateral wrist, then it arborizes about the dorsum of the hand.

Blood vessels

Profunda Brachii artery (deep artery of the arm) a branch from the brachial artery that runs with the radial nerve in the spiral groove on the humerus. It terminates as the middle and radial collateral arteries.

Venae Commitantes these are many small, paired veins that accompany the brachial artery

Dorsal venous arch

Posterior Interosseous Artery a branch from the common interosseous artery of the ulnar artery, easily seen in the deep forearm compartment running with the posterior interosseous nerve.

Cephalic vein located on the lateral border of the wrist and will be connected to the basilic vein via the median cubital vein.

Basilic vein located on the medial border of the wrist and will be connected to the basilic vein via the median cubital vein.

Median cubital vein the connecting vein between Basilic and Cephalic veins. It may or may not be obvious.

6. *Clinical Applications*

a. **Triangular Interval Syndrome** compression of the radial nerve and/or brachial profundus artery within the triangular interval secondary to trauma or muscle hypertrophy.

b. **Crutch Palsy or Saturday Night Palsy** a compressive neuropathy of the radial nerve either in the axilla or upper arm resulting in motor and sensory deficits.

c. **Posterior Interosseous Nerve Syndrome** compression of the posterior interosseous nerve in the forearm resulting in only motor deficits.

d. **Boxer's Fracture** a fracture of the neck of the 5th metacarpal due to trauma.

e. **Boutonniere Deformity** injury to the central slip of the extensor tendon at the base of the middle phalanx resulting in PIP flexion and DIP extension.

f. **Bennett's Fracture** fracture of the base of the first metacarpal often from forced abduction.

C. Pectoral Region

Note: The Pectoral Region must be dissected **prior** to the Axillary Region.

1. *Preparation*

Begin your preparatory work by studying in an atlas the sternum, ribs and clavicle. Also, review the associated blood vessels in the region as well as the nerves of the brachial plexus. Be sure to have an understanding of the nerve fibers carried in each of the nerves of the plexus throughout this dissection.

2. *Cadaver Position*

Place your cadaver in the supine position (face up) with a block under the thorax and one under the posterior neck to keep the head off of the dissection table.

3. Pre-dissection Discussion

First, if you have a female cadaver you will need to remove the breasts then continue with the rest of the pectoral dissection. See below for dissection instructions on how to remove the breasts.

Breast Removal

Make an incision around the entire breast itself but stay superficial to the pectoral fascia that covers the pectoralis muscle. Place your hand behind the breast into the *retromammary space* (see below), and slowly lift and separate it from the underlying fascia. You are tearing the extensions of the suspensory ligaments of the breast, also known as Cooper's ligaments from the pectoral fascia. Be sure to lift and snip as necessary to fully extract the breast from the pectoral region without damaging the pectoral fascia. You may discard the breast unless it is to be studied later. Ask your course instructor for further instructions.

 The skin can be thin in the upper chest, so go slowly. You will remove the skin from medial to lateral but leave the skin attached to use as cover. The platysma is just deep to the skin in the upper chest, so try to save it as well as preserve the Supraclavicular nerves just deep to the platysma. When you clean the pectoralis major you will notice its two heads: the clavicular and the sternocostal head. The deltopectoral groove will be found between the clavicular head and the deltoid muscle and it contains the cephalic vein. You will see the Lateral Pectoral Nerve piercing the clavicular head of the pectoralis major and the Medial Pectoral Nerve piercing the sternal head and the pectoralis minor. Remove the pectoralis major from the sternum and clavicle and reflect it superolaterally, then reflect the pectoralis minor superolaterally. The thoracoacromial trunk will be found on the surface of the pectoralis minor. The lateral thoracic artery is found at the lateral border of the pectoralis minor, and usually terminates at the 5th intercostal space.

For Reference:

The pectoralis minor muscle divides the axillary artery into three parts described as a 1st part, a 2nd part, and a 3rd part. This fact is useful when studying the three parts of the axillary artery, since each part presents with:

One branch from the first part (A)
Two branches from the second part (B)

Three branches from the third part (C)

A. Superior or Highest thoracic artery

B. Thoracoacromial trunk
   Lateral thoracic artery

C. Subscapular artery
   Anterior humeral circumflex
   Posterior humeral circumflex

Note: Additional details of these branches will be discussed in Axillary dissection.

4. Incisions to Make

Make an incision from the jugular notch all the way inferior to just superior to the xiphoid. Then make bilateral incisions along the clavicular margin to the mid-brachial region. Also, make cuts along the midaxillary line to about the costal margin and connect them to the xiphoid on an angle. Finally, cut about the nipples and make an incision from each circular cut across to the other. Removing the deep fascia from the pectoralis major allows you to easily outline its borders and notice its two heads. Cut the pectoralis major at its medial attachments (keeping its lateral attachment intact) but be sure to use blunt dissection to clearly separate it out. As you "flip" it laterally, notice the lateral pectoral nerve and the thoracoacromial artery. Now, notice the clavipectoral fascia covering the pectoralis minor. Reflecting this muscle from its rib attachments allows you to see the medial pectoral nerve piercing this muscle.

5. Structures to Clean and Identify (See Figure 8)

Muscles

Platysma a "paper thin" muscle within the subcutaneous tissue of the neck. It attaches to the pectoral fascia and the mandible but is not considered a muscle of mastication.

Pectoralis Major this muscle presents both a sternocostal head and a clavicular head.

Pectoralis Minor serves as an accessory muscle of respiration in addition to dividing the axillary artery into three parts.

Subclavius a tiny muscle located deep to the clavicle that serves to depress the clavicle and may protect the subclavian vascular structures.

Serratus Anterior a "saw-toothed" muscle that is known as the "boxer's muscle" since it functions to keep the scapula on the thoracic wall during punching maneuvers.

Fascia

Clavipectoral Fascia also known as the costocoracoid membrane, it covers the pectoralis minor and is found between it and the subclavius muscle.

Suspensory ligament of the Axilla a continuation of the clavipectoral fascia within the axilla attaching to the axillary fascia.

Spaces

Deltopectoral triangle a triangular space between the deltoid and the pectoralis major muscles that houses the cephalic vein and the deltoid branch of the thoracoacromial trunk.

Retromammary space a space between the deep layer of superficial fascia and the pectoral fascia

Nerves

Supraclavicular nerves (C3–C4) found just deep to the platysma and described as medial, intermediate, and lateral.

Medial Pectoral nerve (C8–T1) pierces the pectoralis minor and the sternocostal head of the Pectoralis Major. It is from the medial cord even though it is positioned lateral to the lateral pectoral nerve.

Lateral Pectoral nerve (C5–C7) pierces the clavicular head of the Pectoralis Major and it is from the lateral cord but is positioned medial to the medial pectoral nerve.

Long Thoracic nerve (C5–C7 roots) located along mid-axillary line on the anterior surface of the serratus anterior.

Blood vessels

Cephalic vein a fairly obvious vein that is found in the deltopectoral triangle and terminates in the axillary vein.

Superior Thoracic artery a small branch of the first part of the axillary artery that irrigates the first and second intercostal spaces.

Thoracoacromial Trunk these branches are named by where they are headed.

a. clavicular branch

b. pectoral branch

c. acromial branch

d. deltoid branch

A helpful mnemonic–**CADP** (**C**adavers **A**re **D**ead **P**eople)

Lateral Thoracic artery runs along the lateral border of the pectoralis minor. It runs for a short time proximally with the long thoracic nerve.

6. Clinical Applications

a. **Winging of the Scapula** a condition resulting from damage to the long thoracic nerve or to the serratus anterior muscle whereby the scapula rises off of the thoracic cage and appears as a "wing" in pushing maneuvers.

b. **Pectus Excavatum** (funnel chest) a chest wall deformity resulting in a concave depression of the sternum.

c. **Pectus Carinatum** (pigeon chest) the anterior protrusion of the sternum, much less common than funnel chest.

d. **Poland Syndrome** unilateral congenital absence of the pectoralis major and minor muscle with other associated musculoskeletal abnormalities.

D. Axillary Region

1. Preparation

Begin your preparatory work by studying in an atlas the structures that make up the borders of the axilla. It is a pyramidal shaped area between the upper arm and side of the chest. Also, spend some time reviewing the brachial plexus, its location, its formation, and its branches.

2. Cadaver Position

Place your cadaver in the supine position (face up) with a block under the thorax and one block under the posterior neck to keep the head off of the dissection table. You will have to abduct the arm and tie it in place to complete this dissection. Do not abduct the arm so far as to dislocate or fracture the humerus.

3. Pre-dissection Discussion

Strip away the axillary sheath to uncover the axillary vessels and brachial plexus. You should be able to see the three cords of the brachial plexus (medial, lateral and posterior). They are named for their position relative to the axillary artery. You will not see the roots, trunks or divisions clearly. There is a lot of fat in the axillary region, and it will take time to gingerly remove it without destroying the collateral branches of the brachial plexus. Generally speaking, once this area is free of skin and adipose tissue, blunt dissection and the scissors technique are the best methods to clean the dissection field.

4. Incisions to Make

This area may be open already. If not, remove the skin and axillary fascia to expose the deeper structures.

5. Structures to Clean and Identify (See Figure 9)

Muscles

Pectoralis minor this muscle serves as the divider of the axillary artery into its three parts.

Coracobrachialis this muscle often blends with the short head of the biceps brachii. The musculocutaneous (C5–C6) nerve pierces it.

Serratus anterior already discussed.

Connective Tissue

Axillary Sheath encases the axillary artery and vein and the cords of the brachial plexus.

For reference:

Brachial Plexus

1. *Dorsal Scapular nerve* (C5 root)
2. *Nerve to Subclavius* (C5–C6 upper trunk)

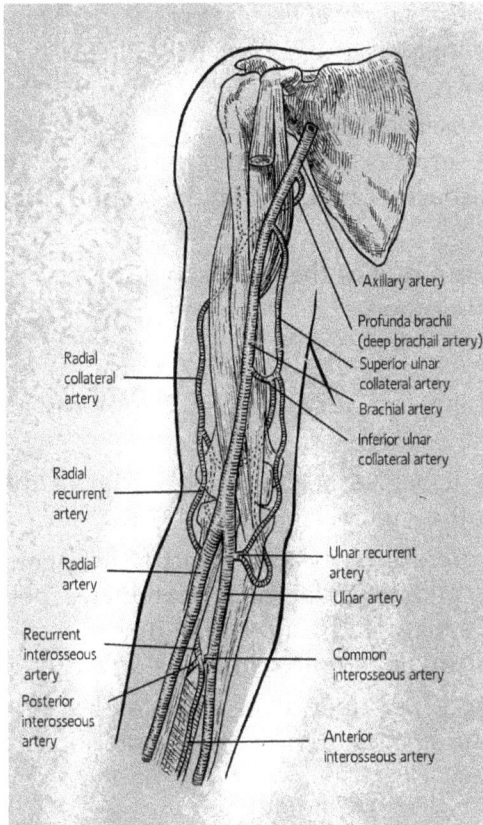

Figure 9. Ventral arm showing brachial artery branches.

3. *Suprascapular nerve* (C5–C6 upper trunk)
4. *Long Thoracic nerve* (C5, C6, C7 roots)
5. *Lateral Pectoral nerve* (C5, C6, C7 Lateral cord)
6. *Upper Subscapular nerve* (C5, C6 Posterior cord)
7. *Middle Subscapular nerve* (C6, C7, C8 Posterior cord)
8. *Lower Subscapular* (C5, C6 Posterior cord)
9. *Musculocutaneous nerve* (C5, C6, C7 Lateral cord)
10. *Axillary nerve* (C5, C6 Posterior cord)
11. *Radial nerve* (C5-T1 Posterior cord)
12. *Ulnar nerve* (C8, T1 Medial cord)
13. *Medial Pectoral nerve* (C8, T1 Medial cord)

14. *Medial Cutaneous nerve of the forearm* (C8, T1 Medial cord)
15. *Medial Cutaneous nerve of the arm* (C8, T1, T2)
16. *Median nerve* (C6-T1)
17. ♦ *Intercostobrachial* (T2)
    pierces the serratus anterior and is often seen "jumping" from 2nd intercostal space to axilla to join the medial cutaneous nerve of the arm.

♦ This nerve is not part of the plexus, but it is typically uncovered during this dissection.

Blood vessels

Axillary vein this vein is very loose and floppy and may be removed to clear the field.

Axillary artery it begins at the outer border of the first rib as a continuation of the subclavian artery.

Superior thoracic artery located about two fingers width below the clavicle at midclavicular line. Terminates between 1st and 2nd intercostal spaces and runs on the upper border of pectoralis minor. It is very small and often breaks very easily.

Lateral thoracic artery found on the lateral aspect of pectoralis minor and on the surface of the serratus anterior.

Subscapular artery

Thoracodorsal
Circumflex scapular

Posterior humeral circumflex artery larger and runs posteriorly around the surgical neck of the humerus entering the quadrangular space.

Anterior humeral circumflex artery smaller and runs deep to the short head of the biceps brachii and coracobrachialis, anterior to the surgical neck of the humerus, and can be located either distal to, or proximal to, posterior humeral circumflex artery.

Note: The Anterior and Posterior Humeral Circumflex arteries *may* have a common trunk from the axillary artery rather than coming off separately.

6. *Clinical Applications*

a. **Referred Pain of a myocardial infarction** The Intercostobrachial nerve (T2) is known to be associated with transmitting pain from the ischemic heart muscle during a myocardial infarction as coming from the axilla and medial aspect of the arm.

b. **Thoracic Outlet Syndrome** (*TOS*) a collective group of signs and symptoms of the upper extremity due to compression of the brachial plexus or the subclavian artery as they traverse the superior thoracic aperture.

c. **Klumpke's palsy** injury to the lower trunk of the brachial plexus (C8-T1) resulting in damage to the median and ulnar nerves.

d. **Erb-Duchenne palsy** *"Waiter's tip" deformity* injury to the upper trunk of the brachial plexus (C5–C6) resulting in damage to the suprascapular, axillary and musculocutaneous nerves.

e. **Carpal Tunnel Syndrome** (*CTS*) compression of the median nerve as it traverses the carpal tunnel resulting in sensory and motor deficits of the thumb and hand. It is classified as a repetitive stress injury.

f. **Pronator Teres syndrome** compression of the median nerve as it passes between the humeral and ulnar head of the pronator teres muscle resulting in significant hand weakness.

g. **Roos' Test** in the seated position the patient positions both arms at 90 degrees and abducts and externally rotates them. The patient then opens and closes the fists slowly for approximately 3 minutes. If the symptoms of thoracic outlet syndrome reappear, they most likely have the syndrome.

h. **Bakody Sign** while seated, the patient places the palm of the affected extremity on top of the head, thus elevating the suprascapular nerve. If the pain is relieved it is radicular, if it exacerbates it the pain is thoracic outlet syndrome.

i. **Allen Maneuver** the examiner flexes the patient's elbow to 90 degrees while abducting and externally rotating the arm as the patient rotates their head to the opposite side. The examiner palpates the ipsilateral radial artery simultaneously. If the pulse disappears the test is positive for thoracic outlet syndrome.

E. Anterior Arm, Anterior Forearm, Cubital Fossa, Anterior Carpus

1. Preparation

Begin your preparatory work by studying in an atlas or on a model the bones of the upper extremity and include in your study the eight carpal bones. Also study the muscles, nerves and connective tissue found here, especially the palmar aponeurosis.

2. Cadaver Position

Supine with a block under the thorax and one under the posterior neck. Forearm will be supinated, and you will have to tie the upper extremity into the proper position to do this dissection.

3. Pre-dissection Discussion

At this point, many structures may have been uncovered and made loose from their connective tissue during a previous dissection. Now use blunt dissection to separate the structures while cleaning the dissection field. Notice in the dissection that muscles are in appreciable layers, superficial to deep. In the cubital fossa notice from lateral to medial, biceps brachii tendon–brachial artery–and the median nerve. As you separate the muscles of the anterior forearm compartment, track their tendons to the Carpal Tunnel.

For Reference:
Carpal Tunnel contents:

> Median nerve
> Flexor Pollicis Longus (1 tendon)
> Flexor Digitorum Profundus (4 tendons)
> Flexor Digitorum Superficialis (4 tendons)

Helpful mnemonic: The "**UP UP**" *does not* enter the carpal tunnel. Those being **U**lnar nerve, **P**almaris Longus tendon, **U**lnar artery, and **P**almar cutaneous branch of the median nerve, which is given off just proximal to the flexor retinaculum.

Also, you will notice the radial artery lateral to the tendon of the FCR and anterior to the styloid process of the radius and within the Anatomical Snuffbox. It will be traveling with the superficial branch of the radial nerve too (lying lateral to it). Be aware of the

Ulnar artery and Ulnar nerve. They *are within the same connective tissue* and must be separated cleanly and carefully. You may miss the ulnar nerve if you are not aware of this fact. The ulnar artery is found lateral to the ulnar nerve and above the flexor retinaculum further distally. These structures pass through *Guyon's Canal* to enter the palm.

4. Incisions to Make

Incise the skin from acromion to wrist if it has not been done prior. Do not leave skin attached. Remove the subcutaneous tissue and also incise the deep fascia.

5. Structures to Clean and Identify (See Figures 10, 11, 12, 13)

Muscles

Coracobrachialis often stuck to the short head of biceps brachii. The musculocutaneous nerve will pass through this muscle.

Biceps brachii
 long head (located laterally)
 short head (located medially)

Brachialis the "workhorse" of the forearm flexors.

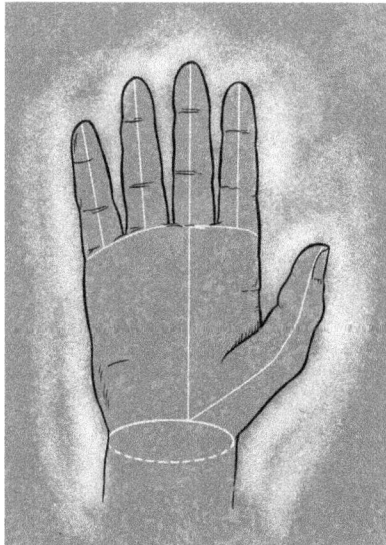

Figure 10. Incisions to make on the palmar aspect of the hand.

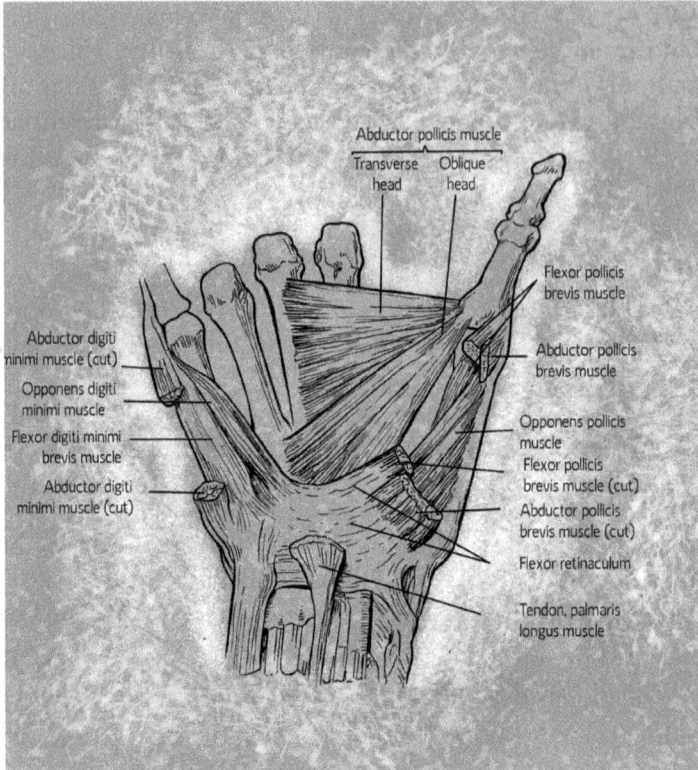

Figure 11. Deep dissection of palmer aspect of hand.

Bicipital aponeurosis aka the "Thanks be to God" aponeurosis because it covers the median nerve and the brachial artery in a protective fashion during venipuncture.

Supinator deep branch of the radial nerve pierces it and then the nerve is known as the *Posterior Interosseous Nerve.*

Brachioradialis the superficial branch of the radial nerve passes deep to it, then arises on the dorsum of the hand to arborize extensively.

Pronator teres
 Superficial head (humeral)
 Deep head (ulnar)

Flexor pollicis longus
Flexor carpi radialis

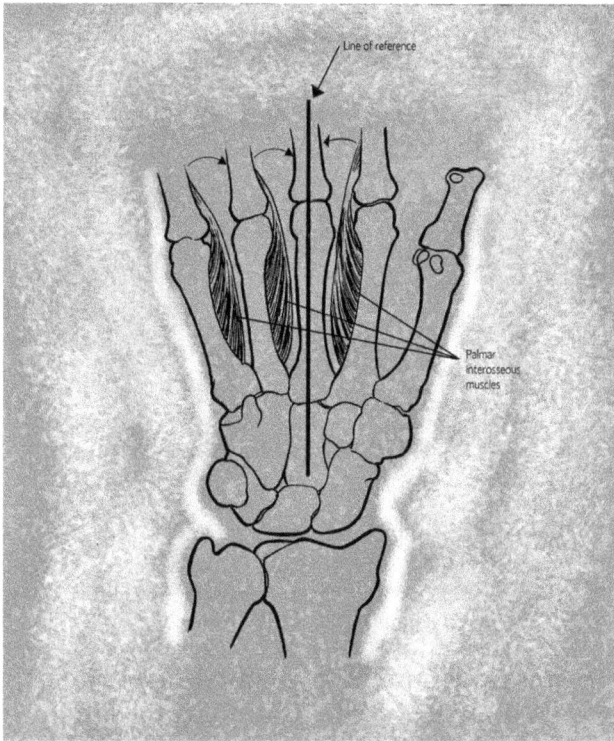

Figure 12. Palmar interossei.

Palmaris longus it may be absent. It does not traverse the carpal tunnel and it attaches to palmar aponeurosis.

Flexor retinaculum (aka transverse carpal ligament) is attached to the hamate and pisiform (medially) and scaphoid and trapezium (laterally).

Flexor carpi ulnaris
Flexor digitorum superficialis

Flexor digitorum profundus innervated by both the median and ulnar nerves.

Pronator quadratus a very deep quadrangular muscle between the distal shafts of the radius and ulna. The anterior interosseous neurovascular bundle will be seen going under it.

Nerves

Median nerve it travels between the humeral and ulnar heads of the pronator teres in the proximal forearm.

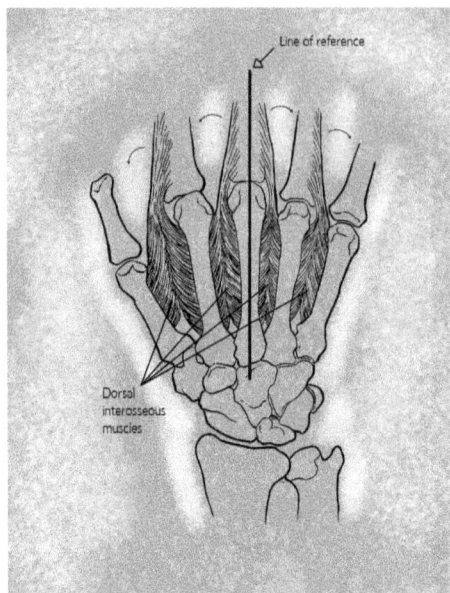

Figure 13. Dorsal interossei.

Anterior Interosseous branch of median nerve it emerges about five centimeters distal to the elbow, running toward the thumb.

Ulnar nerve it passes between two heads of flexor carpi ulnaris.

Deep branch of the radial nerve find it just proximal to the supinator muscle.

Lateral cutaneous nerve of the forearm the terminal portion of the musculocutaneous nerve found passing between biceps brachii and brachialis muscles.

Blood Vessels

Brachial artery arises just after the inferior border of the teres major muscle. It continues into the cubital fossa to divide into the radial and ulnar arteries.

Brachial profundus artery travels with radial nerve in spiral groove on the shaft of the humerus between long and lateral heads of triceps brachii muscle.

Cephalic vein easily identified on the radial side of the forearm.

Basilic vein not so easily identified on the ulnar side of the forearm.

Brachial veins (venae commitantes) this is the collective term for the many brachial veins seen here.

Median cubital vein the connecting vein between the cephalic and basilic veins. A favorite target of the phlebotomist.

Radial artery

a. Radial recurrent artery

Ulnar artery

a. Anterior recurrent branch

b. Posterior recurrent branch

c. Common Interosseous artery
   anterior interosseous artery
   posterior interosseous artery

6. *Clinical Applications*

a. **FOOSHA** (*Fall On the OutStretched HAnd*) an acronym that describes the most common etiology of hand and wrist injuries.

b. **Wartenberg's Syndrome** compression of the superficial branch of the radial nerve at the wrist, often from tight watches resulting in paresthesia of parts of thenar eminence and dorsolateral hand.

c. **Allen's test** a test to suggest ulnar or radial artery occlusion. The examiner occludes and subsequently releases each artery and evaluates the filling time of the hand. If circulation fails to return within five seconds or less it represents vascular compromise.

d. **Phalen's sign** the patient flexes their wrist, and the dorsal surfaces are pressed together and maintained for 60 seconds. A positive sign for carpal tunnel syndrome is indicated by numbness and tingling in the thumb, index finger, middle finger, and lateral half of the ring finger.

e. **Tinel's sign at the wrist** tapping of the carpal tunnel at the wrist. Numbness and tingling of the thumb, index finger, middle finger and lateral half of the ring finger is positive for carpal tunnel syndrome or median nerve irritation.

f. **Colles' fracture** a fracture of the distal radius from FOOSHA (Falling On the OutStretched Hand) resulting in dorsal angulation of the distal fragment seen on plain film.

g. **Smith fracture** a fracture of the distal end of the radius from trauma to the wrist resulting in volar angulation of the distal fragment seen on plain X-ray film.

h. **Ganglion cyst** (*Bible cyst*) variable sized synovial cyst seen at the wrist thought to result from repetitive stress injury, especially noted in young gymnasts.

F. Palmar Aspect of Hand

1. Preparation

Begin your preparatory work by studying the bones (carpals) of the wrist and hand (metacarpals). Also study the muscles, nerves and connective tissue found here. Pay *special attention* to the thenar and hypothenar muscles and their innervations.

2. Cadaver Position

Supine with a block under the thorax and posterior neck. Palm is to be supinated. You will most likely have to tie the hand down to keep it in the proper position.

3. Pre-dissection Discussion

After removing the skin and making the appropriate cuts into the digits and severing the flexor retinaculum, identify the palmar aponeurosis. It is very thick and extremely tough, and as a result, it is *very* difficult to dissect. It will take work to remove but go slowly and it will be worth it. The palmar aponeurosis does provide attachment for palmaris longus and palmaris brevis muscles. Once the aponeurosis is removed, it is easier to see the muscles in the thenar, hypothenar and deep palmar areas, as well as the palmar and digital arteries and nerves.

4. Incisions to Make

Remove the palmar aponeurosis using blunt dissection and occasional scissors as needed to cut the connections (septa) that attach to the metacarpals. Then, remove the skin and superficial fascia from the digits distally. You will eventually need to section the Abductor Pollicis brevis in the middle of its belly in order to clearly see the Opponens Pollicis, which lies deep to it. Be sure to preserve the recurrent branch of the median nerve **(See Figure 10).**

5. *Structures to Clean and Identify* (See Figure 11, Figure 12 and Figure 13)

Muscles

Palmar aponeurosis (and its digital extensions)

Abductor pollicis brevis the most superficial muscle of the thenar eminence.

Flexor pollicis brevis

a. Superficial head

b. Deep head

Opponens pollicis
Tendon of flexor pollicis longus
Adductor pollicis

a. Oblique head

b. Transverse heads

Abductor digiti minimi
Flexor digiti minimi
Opponens digiti minimi

Palmaris brevis this is often within the palmar aponeurosis in the hypothenar eminence and the fibers seem to be scattered and run transversely across the palm.

Lumbricals these are "worm-shaped" muscles that attach directly to the four tendons of the flexor digitorum profundus muscle.

Interossei these are between the metacarpal shafts and there are three palmar and four dorsal interossei.

Note: **P**almar **Ad**duct and **D**orsal **Ab**duct the digits.

Nerves
Median nerve

a. Palmar cutaneous branch travels *above* the transverse carpal ligament

b. Recurrent branch this small but mighty nerve heads directly into the abductor pollicis brevis in a "candy-cane" shaped trajectory

c. Palmar digital sensory branches

Ulnar nerve
a. Superficial branch and its palmar digital branches
b. Deep branch

Blood Vessels
Ulnar artery
a. Superficial palmar arch this branch helps to form the superficial palmar arch.
b. Deep palmar branch this branch helps to form the superficial and deep palmar arterial arches along with the radial artery and its superficial and deep branches.
c. Common digital arteries
d. Proper digital arteries

Princeps pollicis from radial artery located between first dorsal interossei and the oblique head of the adductor pollicis.

Radial artery
a. Superficial branch
b. Deep palmar branch the deep branch helps to form the deep palmar arch.

6. *Clinical Applications*

a. **Dupuytren's Contracture** a painless thickening of the tissue in the hand and fingers that causes the inability to straighten the fingers or flatten the hand. It especially affects the ring and little fingers.
b. **Mallet Finger** caused by forced hyperflexion of the distal interphalangeal joint, resulting in avulsion of the attachment of the extensor tendon from the base of the distal phalanx.
c. **Trigger Finger** tenosynovitis involving the flexor digitorum superficialis. The tendon becomes thick and inflamed and can no longer pass through the sheath causing it to "catch and snap" resulting in a painful nodule.
d. **Swan Neck Deformity** injury to extensor tendon at the DIP resulting in hyperextension of PIP and flexion of DIP.

e. **Claw hand** (*distal ulnar n. injury*) abnormal position of the hand due to injury of the ulnar nerve, most often from lower ulnar nerve lesion.

f. **Ape Hand** (*median n. injury*) injury to the median nerve, especially the recurrent branch, resulting in the inability to oppose the strong adductor pollicis muscle positioning the thumb in the same plane as all the other digits, resembling the hand of lower primates.

g. **Hand of Benediction** (*median n. injury*) injury to the median nerve, especially proximally, resulting in the inability to make a fist when asked.

h. **Froment's Sign** the patient is asked to grab a piece of paper between the thumb and index finger while the examiner tries to pull it away. Inability to hold on to the paper indicates ulnar nerve injury because of the weakness of the adductor pollicis muscle.

i. **Gamekeeper's Thumb** a rupture of the ulnar collateral ligament at the 1st metacarpophalangeal joint.

UNIT 3
Thorax and Mediastinum

A. Lungs and Pleural Cavity

1. Preparation

Begin your preparatory work by studying the bones of the thorax, including the ribs, sternum, clavicle and vertebrae. Also, review the mediastinum, its divisions, and its contents.

2. Cadaver Position

Supine with a block placed under the head and a block under the upper thorax.

3. Pre-dissection Discussion

At this point, the chest plate will be removed. In this dissection you will remove and dissect the thoracic viscera. The heart and lungs will be removed and discussed separately. The thymus gland will most likely be involuted and not easily seen. You will need to identify and study the sternal angle of Louis and its clinical relevance. Also, preview the mediastinal divisions and their contents, including blood vessels and nerve trajectories.

For Reference:

Superior Mediastinum (the Angle of Louis marks the inferior border)

  a. *thymus gland*

  b. *arch of aorta*

  c. *superior vena cava*

  d. *pulmonary trunk*

e. *trachea and primary bronchus*

f. *esophagus* the diaphragm marks the inferior border and the pericardium divides the inferior mediastinum into three parts.

1. Anterior Mediastinum (between sternum and pericardium)

   a. *thymus gland-inferior part*

   b. *sternopericardial ligaments* ("cardiac seatbelts")

2. Middle Mediastinum

   a. *pericardial cavity with heart*

   b. *phrenic nerves and root of the lung*

3. Posterior Mediastinum (between pericardium and T5–T12)

   a. *esophagus*

   b. *azygos vein*

   c. *hemiazygos vein*

   d. *accessory hemiazygos vein*

   e. *thoracic aorta*

   f. *thoracic duct*

   g. *vagus nerves*

   h. *thoracic aorta*

   i. *central tendon of the diaphragm*

   j. *posterior intercostal arteries*

   k. *sympathetic trunk and paravertebral ganglia*

   l. *Splanchnic nerves*
      Greater splanchnic (T5–T9)
      Lesser splanchnic (T10–T11)
      Least splanchnic (T12)

4. Incisions to Make

Cut the clavicles at their midpoint and cut ribs 1–6 about 1–1.5 inches posterior to the mid-axillary line. Cutting posterior to the mid-axillary line will give you a lot more room to work within the thorax. If you wish to remove the plate entirely and not keep it attached, you will need to lift and snip many structures attached to the sternum such as muscles, blood vessels, the sternopericardial ligaments and the sternal attachment of the SCM. It is usually preferred to remove the chest plate entirely.

5. *Structures to Clean and Identify* (See **Figure 14 and Figure 15)**

Muscles

Subclavius

Intercostal muscles
external
internal
innermost

Serratus anterior

Transversus thoracis found on the inside of the chest plate

Diaphragm major muscle of respiration and is covered by diaphragmatic pleura.

Connective tissue

Endothoracic fascia a fine membrane superficial to parietal pleura and thickens up on the apex of the lung to become the suprapleural membrane.

Pleura
Parietal (lines the wall of a cavity)
mediastinal
diaphragmatic
cervical
costal
Visceral (covers the organ itself)

Sternopericardial ligaments aka "cardiac seatbelts" very tough thickenings of the fibrous pericardium that anchor the heart to the posterior sternum.

Ligamentum arteriosum a remnant of the fetal ductus arteriosus and it connects the left pulmonary artery to the arch of the aorta.

Spaces

Mediastinum described as the space between the pleural cavities with divisions and very specific contents.

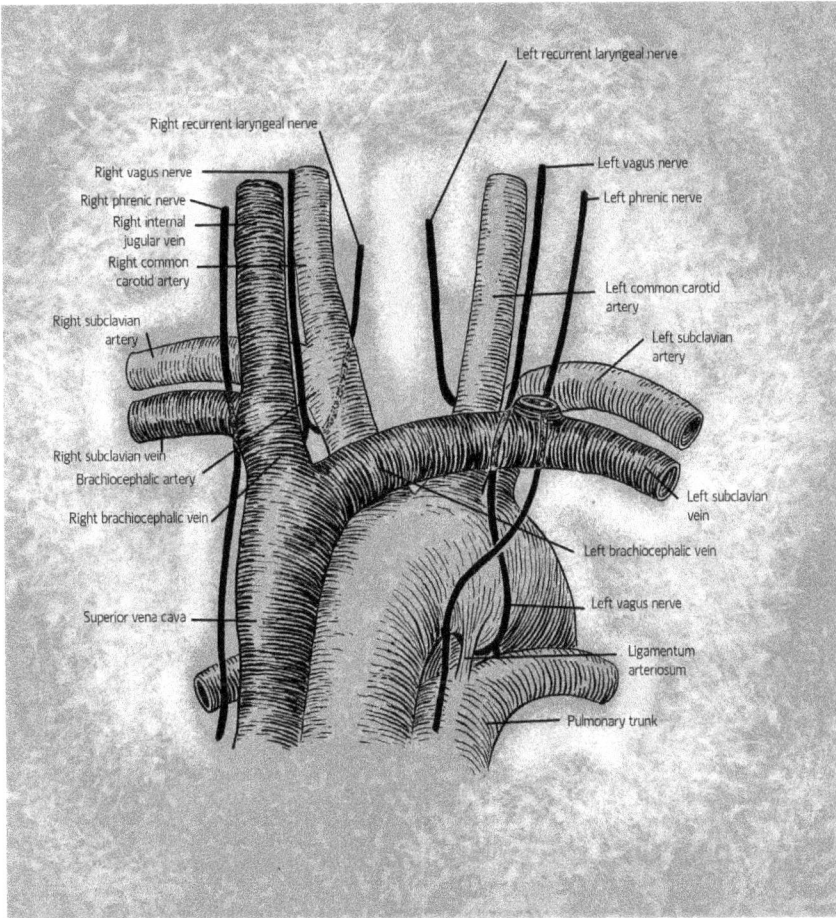

Figure 14. Superior mediastinum contents.

Costodiaphragmatic recess located where the costal pleura and diaphragmatic pleura meet.

Costomediastinal recess located posterior to the sternum where the costal pleura meets mediastinal pleura.

Esophageal hiatus an opening in the diaphragm for the esophagus, vagal trunks and esophageal branches of left gastric artery.

Caval hiatus an opening in the diaphragm for the inferior vena cava and right phrenic nerve.

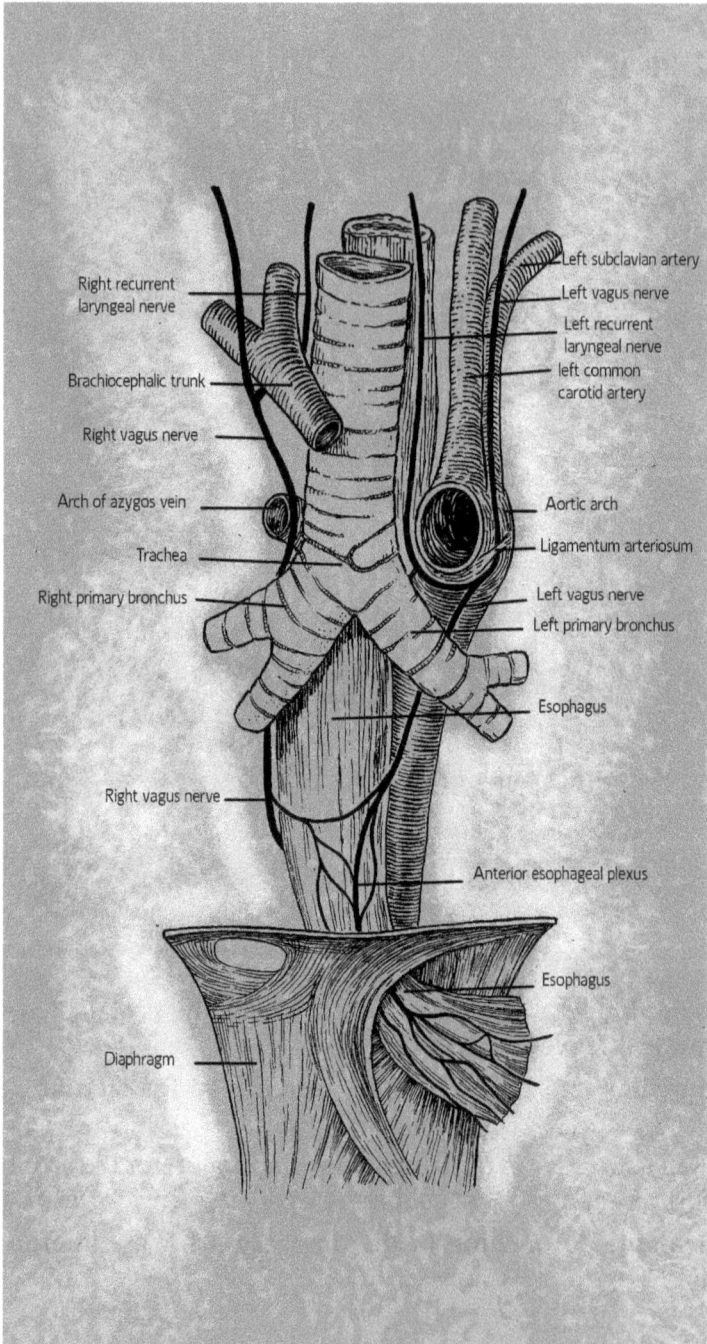

Figure 15. Nerves of mediastinum. Heart and lungs have been removed.

Aortic hiatus an opening (not a true hiatus) in the diaphragm for the aorta, azygos vein and thoracic duct. The thoracic duct is located between the thoracic aorta and the azygos vein. Also, the aortic hiatus is formed by the right and left crura of the diaphragm.

Nerves

Phrenic nerve C3, **C4,** C5-"Keeps the diaphragm alive" and is easily seen on anterior scalene.

Vagus nerve

Recurrent laryngeal a sensory and motor nerve from the vagus to most of the muscles of the larynx.

Intercostal nerves they are the ventral rami of the thoracic nerves.

Sympathetic trunk (sympathetic chain)

Splanchnic nerves
   Greater (T5–T9)
   Lesser (T10–T11)
   Least (T12)

♦ Collectively, these are preganglionic sympathetic fibers that pass through the sympathetic chain without synapsing, but do eventually synapse on the prevertebral ganglia like the celiac ganglion, superior and inferior mesenteric ganglion, etc.

Blood Vessels

Azygos vein found on the right side of the thoracic vertebral bodies helping to drain the posterior walls of the thorax and abdomen. It arches superiorly and drains into the superior vena cava.

Hemiazygos vein located on the left side of the thoracic vertebral bodies and crosses the midline to join the azygos vein at approximately the T9 level.

Accessory hemiazygos vein found at about the level of the 4th or 5th intercostal space on the left upper thoracic region and crosses over to join the azygos vein at approximately the T7 vertebral level.

Posterior intercostal arteries the 1st and 2nd posterior intercostal arteries arise from the supreme thoracic artery while the remainder arise from the thoracic aorta.

Thoracic aorta begins just after the arch of the aorta at the Angle of Luis and continues through the aortic hiatus where it then becomes the abdominal aorta.

Anterior Intercostal arteries the upper six are from the internal thoracic while the lower seventh to the ninth are from the musculophrenic artery.

Internal thoracic (internal mammary) artery arises from the subclavian on the inside of the chest plate lateral to the sternum and anterior to the transversus thoracis muscle. It will give rise to several intercostal arteries and perforating arteries. It terminates by giving rise to the superior epigastric and musculophrenic arteries.

Superior epigastric artery *medial* terminal branch of internal thoracic artery. The superior epigastric artery will anastamose with the inferior epigastric artery from the external illiac within the posterior rectus sheath.

Musculophrenic artery *lateral* terminal branch of the internal thoracic artery. The musculophrenic artery will anastamose with the inferior phrenic arteries and the ascending branch of the deep circumflex illiac as well as the tenth and eleventh posterior intercostal arteries too.

Pericardiacophrenic artery found running with the phrenic nerves in the thorax within the same connective tissue. Pericardiacophrenic artery is from the internal thoracic artery.

Brachiocephalic veins from the union of the internal jugular vein and subclavian veins.

Superior Vena Cava (SVC)

Aortic arch gives rise to the brachiocephalic trunk, left common carotid and the left subclavian artery.

Brachiocephalic trunk will divide into the right common carotid artery and the right subclavian artery.

Right Common carotid artery

Right Subclavian artery

Left Common carotid artery

Left Subclavian artery

6. *Clinical Applications*

a. **Thyroid surgery and the recurrent laryngeal nerve** iatrogenic injury of the recurrent laryngeal nerve is one of the main problems in thyroid surgery. Vocal cord paresis or paralysis can occur resulting in hoarseness, dyspnea, and glottal obstruction.

B. Lungs

1. Incisions to Make or Actions to Take to Remove the Lungs

Place your hands behind the lung and gently tug antero-laterally. However, before the lungs become loose you may have to break the many strong adhesions of the visceral pleura to the parietal pleura. Next, use scissors to cut the root of the lung at its midpoint. You are cutting the bronchi, pulmonary artery and pulmonary veins. Gently remove the lung from the thoracic cavity. Do not damage the phrenic nerves or the vagus nerves during removal!

2. Structures to Clean and Identify

Root of the Lung it is the point where various structures enter and exit the lung.

Lobes (superior, inferior and middle)

Fissures (oblique and horizontal)

Surfaces

Lingula "little tongue" (found only on left lung)

Pleura
   cervical
   mediastinal
   costal
   diaphragmatic

Pulmonary ligament is a double fold of pleura running from the hilum to the root, in a sickle-shaped orientation.

Cardiac notch
   left lung-superior lobe

Cardiac impression
   right lung

Grooves for:
   esophagus
   arch of the aorta
   arch of azygos vein
   thoracic aorta

Eparterial bronchus only on right lung found just above pulmonary artery.

3. Clinical Applications

a. **Atelectasis** a term describing a small area of lung collapse.

b. **Pneumothorax** defined as air in the pleural space which can lead to atelectasis.

c. **Pancoast tumor** it is a tumor of the apex of the lung, either primary or secondary often resulting in compression of the brachial plexus and disruption of the sympathetic chain, potentially giving rise to Horner's Syndrome.

C. Heart and Pericardial Cavity

1. Pre-dissection Discussion

When preparing to remove the heart, take special care to properly remove the pericardium. You may discard the pericardium. Also, be very aware of the great vessels and nerves since the pericardium covers the great vessels too. They all need to be preserved for study. Keep the ligamentum arteriosum attached to the left pulmonary artery and arch of aorta on the cadaver. Preserving the phrenic and vagus nerves is important too.

2. Incisions to Make or Actions to Take to Remove the Heart

Use scissors to open the pericardium or pericardial sac.

1. Find the transverse pericardial sinus and place a probe within it then gently slide it as far superiorly as it will allow.
2. Cut the aorta and pulmonary trunk slightly above their exit point from the heart, meaning just below the probe.
3. Cut the SVC just above its entrance into the right atrium

Be sure to preserve the Arch of the Azygos vein here,

4. Cut the IVC very close to the diaphragm.
5. Cut the pulmonary veins as they enter the left atrium.
6. Remove the heart and flush it with water to remove the clots. You will need a blunt probe or hemostats to pick out the clots.

3. *Structures to Clean and Identify (Externally)* (See Figure 16, Figure 17 and Figure 18)

Parietal pericardium

Pericardial cavity or space

Visceral layer of serous pericardium (epicardium)

Myocardium

Transverse pericardial sinus posterior to the ascending aorta and pulmonary trunk.

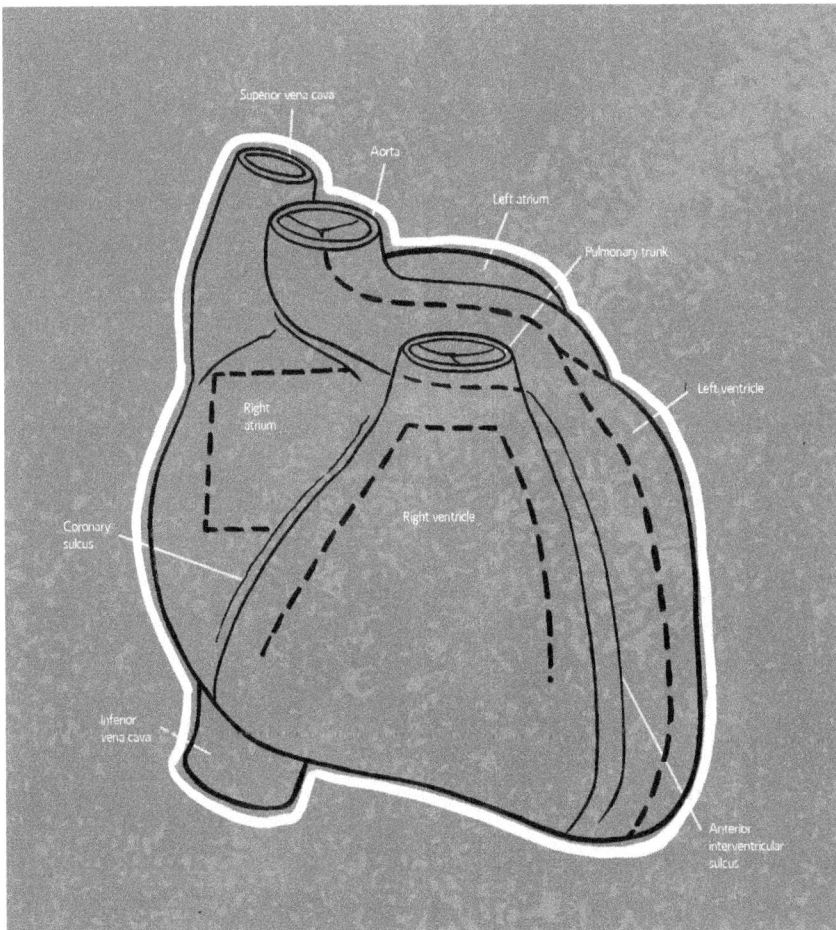

Figure 16. Incisions to make on the heart.

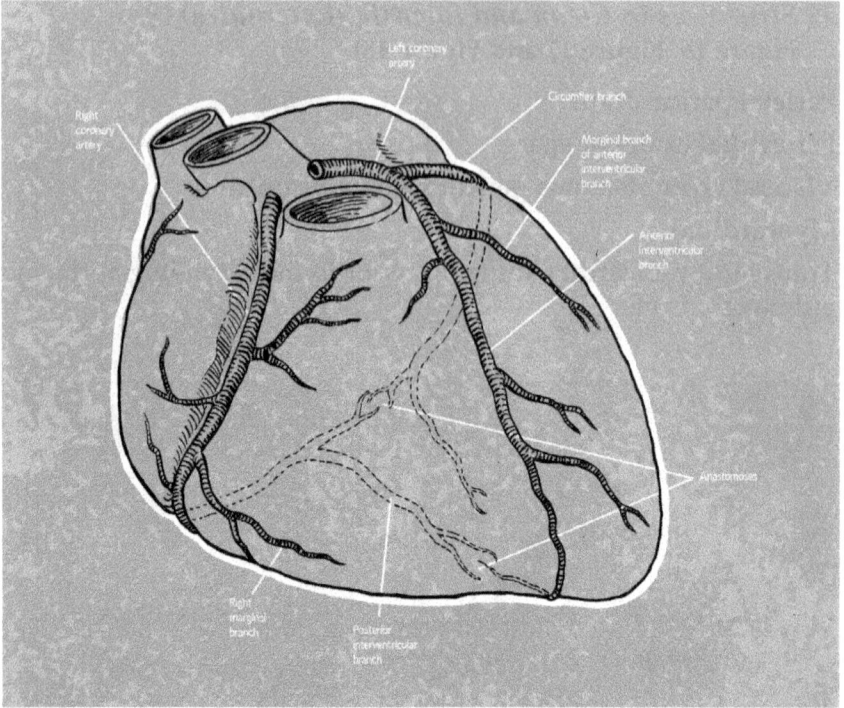

Figure 17. Coronary arteries.

Oblique sinus a "cul-de-sac" in the pericardial cavity posterior to the heart.

Superior Vena Cava (SVC)

Inferior Vena Cava (IVC)

Ascending Aorta

Arch of Aorta

Pulmonary trunk

Ligamentum arteriosum a ligamentous structure connecting the left pulmonary artery to arch of the aorta.

Coronary groove a groove that separates the upper atria from the lower ventricles.

Interventricular grooves

   Anterior groove found on the anterior and houses the anterior interventricular artery and the great cardiac vein.

   Posterior groove found on the posterior and houses the posterior interventricular artery and the middle cardiac vein.

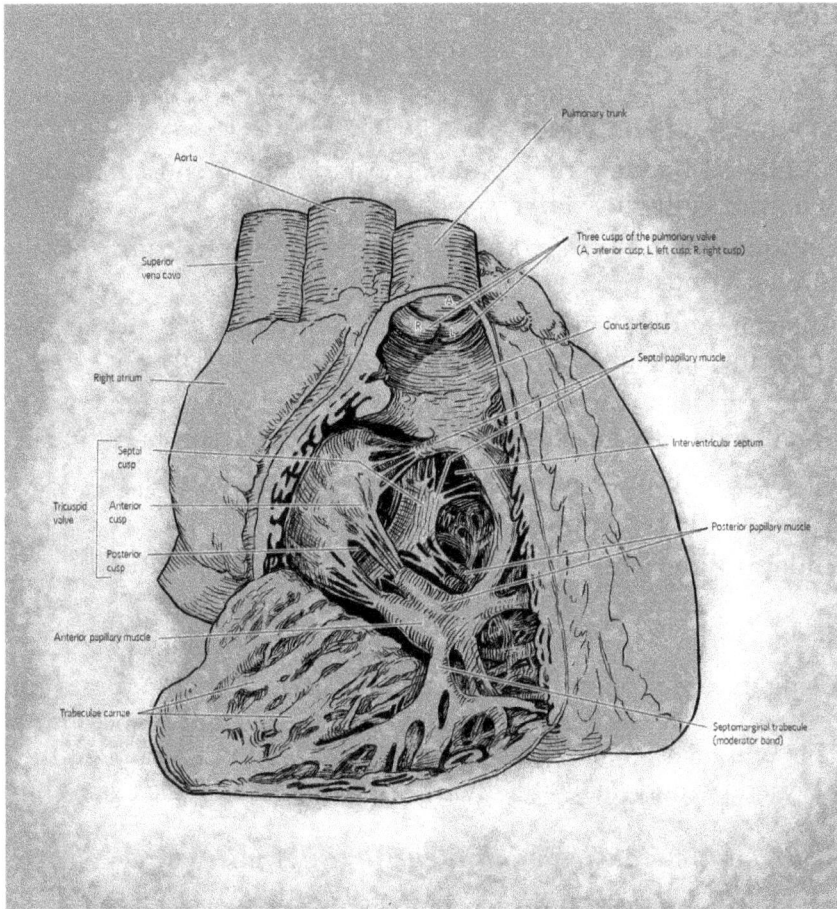

Figure 18. Inside of the heart chambers.

Coronary arteries
Left coronary artery
    Left anterior descending or left interventricular artery
    Circumflex artery
Right coronary artery

a. right marginal

b. posterior interventricular artery

c. conus artery
    sinoatrial nodal branch

Coronary sinus found on the posterior aspect within the coronary groove. It terminates into the right atrium.

Great cardiac vein runs with the anterior interventricular artery and drains into the coronary sinus.

Middle cardiac vein runs with the posterior interventricular artery and drains into the coronary sinus.

Oblique vein (of Marshall) drains the posterior left atrium into the coronary sinus.

Anterior cardiac veins drain directly into the right atrium, just under the right auricle.

Small cardiac veins found running with the right marginal artery draining directly into the right atrium or into the coronary sinus.

Epicardium is the visceral layer of the serous pericardium.

Myocardium is the heart muscle itself.

Endocardium is the smooth endothelial lining of the heart.

4. Incisions to Make

Using tweezers, scissors and a probe, begin by clean the epicardium as best you can to identify the coronary vessels and myocardium. This will take time. Now, open the heart. First open the RA, then RV then LA and finally the LV. See picture showing the cuts.

5. Structures to Clean and Identify (Internally)

Right atrium

Right auricle

Left atrium

Left auricle

Right ventricle

Left ventricle

Pectinate muscles are "comb-like" muscles arising from the crista terminalis.

Crista terminalis

Fossa ovalis an embryological remnant of the foramen ovale on the interatrial septum.

Limbus fossa ovalis a semi-circular ridge about the fossa.

Mitral valve (bicuspid valve)

Tricuspid valve and cusps

Aortic valve and cusps

Pulmonic valve and cusps

Chordae tendineae

Papillary muscles There are 3 papillary muscles in the right ventricle and 2 papillary muscles in the left ventricle.

Interatrial septum

Aortic sinuses small depressions in wall of aorta behind the valves.

Trabeculae Carnea

Septomarginal trabeculae (moderator band) this modified trabeculae carnae transmits the right bundle branch of the electrical system.

Conus arteriosus (infundibulum) smooth outflow tract of the pulmonary trunk.

Supraventricular crest

Sinus venarum

Interventricular septum

Endocardium

6. *Clinical Applications*

a. **Angina Pectoris** described as episodic ischemic cardiac pain not associated with myocardial infarction.

b. **Cardiac Tamponade** an increase in pericardial pressure due to fluid accumulation within the pericardial cavity that interferes with ventricular diastolic filling.

c. **Tetralogy of Fallot** a congenital cyanotic congenital cardiac anomaly characterized by a large ventricular septal defect, stenosis of the pulmonary trunk, an overriding aorta and right ventricular hypertrophy.

d. **Mitral valve Prolapse** (floppy valve syndrome) a degenerative change in the valve leaflets that leads to regurgitation of blood from the left ventricle back into the atrium throughout systole producing a murmur.

UNIT 4
Abdomen

A. Superficial Abdomen

1. Preparation

Begin your preparatory work by studying the bony and non-bony landmarks of the abdominopelvic area, the abdominal quadrants, the abdominal regions, the folds, and the various organs and their positions. Include in your study the inguinal ligament and inguinal canal too.

2. Cadaver Position

Supine with a block placed under the head and a block under the upper thoracic spine.

3. Pre-dissection Discussion

Within the superficial fascia you will find many veins. These include the superficial epigastric, superficial circumflex and thoracoepigastric veins. Generally, these connect the superficial inguinal veins with those about the umbilicus and the intercostal veins. These become important clinically. See clinical applications. Study the composition of the anterior and posteriors layers of the rectus sheath above the umbilicus, at the umbilicus and below the umbilicus. Also study the mesentery of the abdomen.

4. Incisions to Make

Cut from the xiphoid to the pubic symphysis, cutting around the umbilicus. Do not cut directly through the umbilicus. Make two or three cuts from the xiphoid at 45 degree angles to the midaxillary line, then continue to the pubic symphysis. Keep the skin attached to use as a cover to prevent drying out.

5. Structures to Clean and Identify Superficially in the Abdomen (See Figures 19 and 20)

Connective tissue

Camper's fascia superficial fatty layer of the superficial fascia.

Scarpa's fascia deeper membranous layer of superficial fascia just superior to the aponeurosis of external abdominal oblique.

Linea semilunaris (Spigelian Line) lateral border of rectus abdominis muscle.

Linea alba running from pubic symphysis to the xiphoid process.

Tendinous intersections separate muscles of the rectus abdominis.

Arcuate line or semicircular line of Douglas it is the lower border of the posterior rectus sheath. Below the arcuate line is transversalis fascia.

Median umbilical fold obliterated allantois running from the apex of bladder to umbilicus.

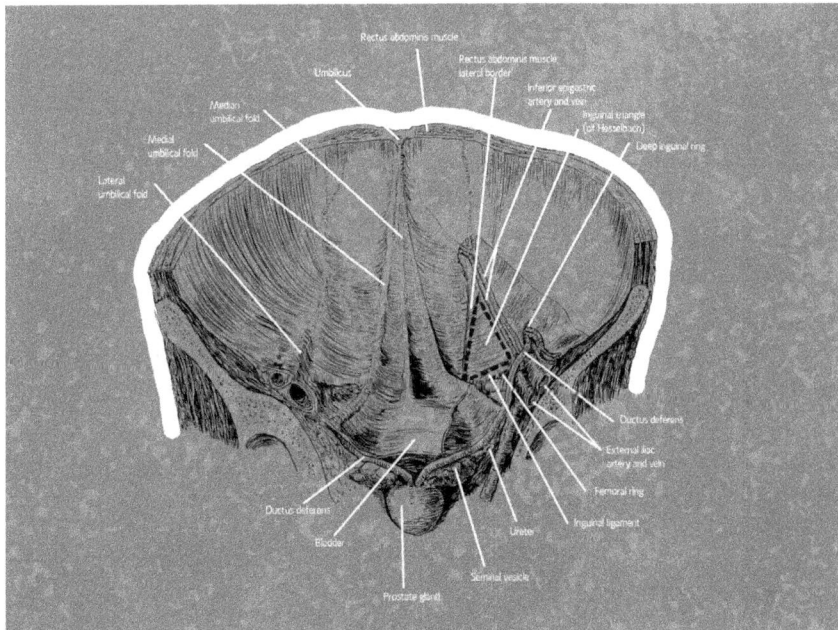

Figure 19. Posterior abdominal wall.

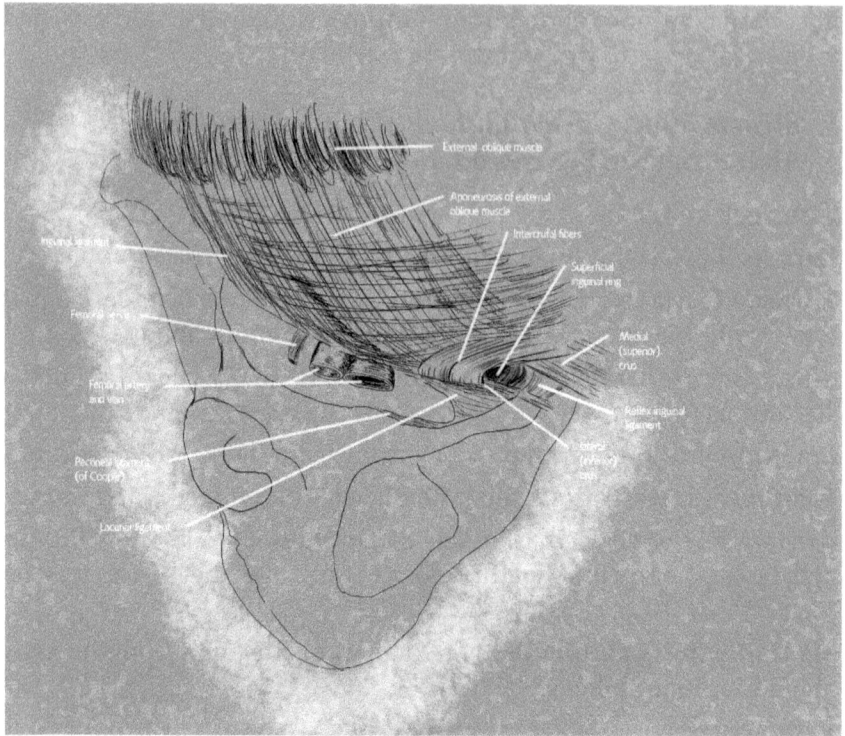

Figure 20. Lower anterior abdominal wall.

Medial umbilical folds obliterated fetal umbilical artery.

Lateral umbilical folds overlies inferior epigastric artery and it is used to classify hernias into either the direct-type or indirect-type.

Transversalis fascia it lines the inner surface of the abdominal muscles and the deep inguinal ring is an opening in this fascia.

Conjoint tendon (falx inguinalis) the combined aponeuroses of internal oblique and transversus abdominis, inserting on the pubic crest forming the medial part of the posterior wall of the inguinal canal.

Greater Omentum aka "Abdominal Policeman" It is a double layer of mesentery folded upon itself, for a total of four layers:

a. splenorenal (lienorenal) ligament

b. gastrosplenic ligament

c. gastrophrenic ligament

d. gastrocolic ligament

Lesser Omentum

a. hepatogastric ligament

b. hepatoduodenal ligament described as the free edge of the lesser omentum contains the hepatic artery, portal vein and bile duct.

Arcuate Ligaments

a. **Median** formed from right and left crura of the diaphragm creating the aortic hiatus.

b. **Medial** arches over psoas major.

c. **Lateral** arches over quadratus lumborum.

Root of the Mesentery origin of the mesentery of the small intestine from the posterior parietal peritoneum attaching to the posterior abdominal wall.

Ligament of Treitz found at the duodenal jejunal flexure. It contains both smooth and skeletal muscle. Anatomically speaking, it separates the foregut from midgut.

Muscles

Rectus abdominis

Pyramidalis muscle may be absent or may be only unilateral.

External abdominal oblique a defect in the fascia of this muscle provides the superficial inguinal ring and the free edge of this aponeurosis forms the inguinal ligament.

Internal abdominal oblique the muscle continues in the spermatic cord to form cremaster muscle.

Transversus abdominis its aponeurosis along with the tendon of the internal abdominal oblique forms the conjoint tendon, aka falx inguinalis.

Psoas major

Psoas minor this muscle may be absent or present unilaterally.

Quadratus lumborum

Spaces

Superficial inguinal ring is a triangular opening in the external abdominal aponeurosis.

Components include:
medial crus (attaches to the pubic crest)
lateral crus (attaches to the pubic tubercle)
intercrural fibers

Hesselbach's Triangle (medial inguinal fossa)
Medial border–lateral border of rectus abdominis muscle
Lateral border–medial border is inferior epigastric vessels
Inferior border–inguinal ligament

Greater Peritoneal Sac

Lesser Peritoneal Sac (omental bursa)

Epiploic foramen (of Winslow) a connection between the greater and lesser sacs.

Boundaries:

a. caudate lobe (superiorly)

b. IVC (posteriorly)

c. hepatoduodenal ligament (anteriorly)

Contents of Inguinal Canal

MALE

Spermatic Cord-contents

a. ductus deferens

b. deferential artery—from superior vesical artery

c. testicular artery—from abdominal aorta

d. cremasteric artery—from inferior epigastric artery

e. pampiniform plexus

f. genital branch of genitofemoral n.

g. sympathetic nerve fibers

h. cremaster muscle

i. lymphatics

Note: The Ilioinguinal nerve enters the canal but *does not* enter through the deep inguinal ring, but it does exit via the superficial inguinal ring!

FEMALE

Round Ligament of the Uterus inserts into the dermis of the labia majora.

Genital branch of genitofemoral n.

Note: The Ilioinguinal nerve enters the canal but does not enter through the deep inguinal ring, but it does exit via the superficial inguinal ring!

Nerves

Anterior cutaneous nerves derived from T7–T11 levels and often referred to as the "thoracoabdominal nerves".

Iliohypogastric nerve penetrates the external oblique fascia just 2 cm superiorlaterally to the superficial inguinal ring. Innervates skin above pubic symphysis and pubic crest.

Ilioinguinal nerve located between internal oblique and transversus abdominis then enters through the inguinal canal and out the superficial inguinal ring to innervate perineal area.

Organs and features

Stomach
Cardia
Fundus
Body
Pylorus
greater curvature
lesser curvature

Gallbladder
body
neck
duct

Liver
porta hepatis
right lobe
left lobe
quadrate lobe
caudate lobe
bare area

Ligaments of liver
a. **falciform ligament** connects liver to anterior AB wall.
b. **ligamentum teres** remnant of the left umbilical vein of the fetus.

c. **coronary ligaments** surrounds the "bare area" of the liver.

d. **triangular ligaments** consists of the points where the parts of the coronary ligaments meet.

e. **ligamentum venosum** a remnant of the ductus venosus that runs from the portal vein to the IVC between the left lobe and caudate lobe of the liver.

Spleen and its surfaces

Small Intestine
duodenum and its parts (first, second, third and fourth)
jejunum
ileum

Colon
cecum
ascending part
hepatic flexure
transverse part
splenic flexure
descending part
sigmoid portion
rectum

Haustra
Tenia coli
Epiploic appendages aka "fat tags"
Transverse mesocolon
Sigmoid mesocolon

Vermiform Appendix (and its mesoappendix)

Pancreas
head
neck
uncinate process
body
tail

Blood Vessels

Superior Epigastric artery a branch from the internal thoracic artery.

Inferior Epigastric artery from the external iliac artery.

Hepatic Portal Vein (HPV) formed from union of superior mesenteric vein and the splenic vein behind the neck of the pancreas.

Proper hepatic artery so named after the common hepatic gives the gastroduodenal artery.

Celiac trunk the artery of the foregut

a. splenic artery (very large and tortuous)
b. common hepatic artery
c. left gastric artery (the smallest of the 3 branches)

Cystic artery usually arises from the right hepatic artery.

Right gastric artery located along lesser curvature and will anastamose with left gastric artery along the lesser curvature of the stomach.

Right gastroepiploic artery arises from the gastroduodenal artery and travels along the greater curvature of the stomach to anastamose with the left gastroepiploic artery.

Left gastroepiploic artery arises from the splenic artery and travels along the greater curvature of the stomach to anastamose with the right gastroepiploic artery.

Superior mesenteric artery

a. inferior pancreaticoduodenal
b. ileocolic
c. right colic
d. middle colic

Superior mesenteric vein will join with the splenic vein to form the hepatic portal vein behind the neck of the pancreas.

Inferior Mesenteric artery

a. left colic
b. sigmoidal arteries
c. superior rectal is the termination of the inferior rectal artery

Inferior mesenteric vein drains usually into the splenic vein.

Marginal artery (of Drummond) this artery connects the terminal branches of superior and inferior mesenteric arteries.

6. *Structures to Clean and Identify in the Deep Abdomen (See Figures 21 and 22)*

Muscles

Psoas major

Psoas minor (may be absent)

Quadratus Lumborum

Iliacus a combination muscle as it is joined by the psoas major to become the Iliopsoas.

Spaces

Deep inguinal ring an opening in the transversalis fascia lying just above the middle of the inguinal ligament.

Nerves

Lumbar Plexus (T12-L4)

Subcostal (T12) found just below the 12th rib

Iliohypogastric (L1)

Ilioinguinal (L1) found between the internal abdominal oblique and transversus abdominis.

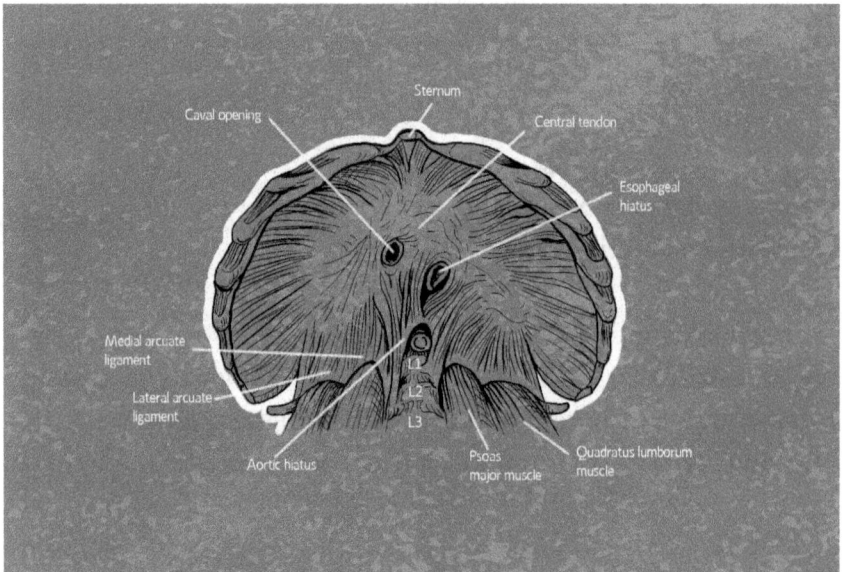

Figure 21. Viewing under the diaphragm to show the various openings.

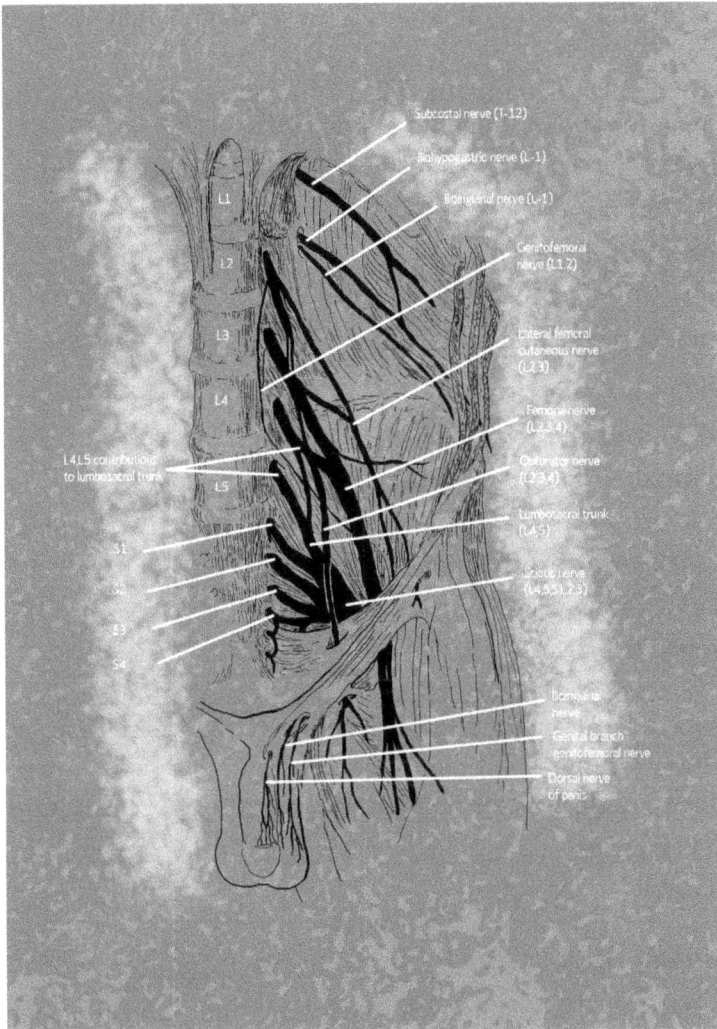

Figure 22. Deep posterior abdominal wall.

Genitofemoral (L1, L2) found passing through the psoas major as it divides into the genital and femoral branches.

a. Genital branch *of genitofemoral* innervates the Cremaster muscle—is the *efferent* limb of Cremasteric reflex.

b. Femoral branch *of genitofemoral* sensory to the anterosuperior thigh and is the *afferent* limb of the cremasteric reflex. Some authors include the illioinguinal nerve as also contributing to the afferent limb of the cremasteric reflex.

Obturator (L2, L3, L4) medial to psoas major—has anterior and posterior divisions.

Lateral Femoral Cutaneous (L2, L3) travels widely along the iliac fossa piercing the inguinal ligament.

Femoral (L2, L3, L4) Located laterally to Psoas Major and is the largest branch of the lumbar plexus.

Organs and Features

Kidney

Adrenal (suprarenal glands)

Ureters

Blood Vessels

Abdominal aorta

Inferior Vena Cava

Testicular (or ovarian) arteries

Testicular (or ovarian) veins

Common Iliac arteries

Deep circumflex illiac artery found running superiorly and posteriorly up along the iliac crest between transversus and internal abdominal oblique.

Inferior Phrenic arteries first pair of arteries under the diaphragm.

Lumbar arteries

Renal artery

7. *Clinical Applications*

a. **Portal-Caval Anastamoses** is collateral communication between the portal and systemic venous system. See examples below. The bolded text is the clinical observable effects of portal-caval backup.

1. *Upper GI tract— Left gastric vein anastamose with esophageal branches of Azygos vein* —**Esophageal varices**

2. *Upper Anal Canal— Superior rectal veins anastamose with Middle and Inferior rectal veins.*—**Internal hemorrhoids**

3. *Umbilical Region—Paraumbilical veins with superficial and deep epigastric veins.*—**Caput Medusae "Head of Medusa".**

b. **Meckel's Diverticulum** incomplete obliteration of the vitelline duct, which connects the midgut to the yolk sac in the fetus. It is usually located around the middle to distal ileum.

c. **Abdominal Aortic Aneurysm** (AAA) an abnormal focal dilation of the abdominal aorta.

d. **McBurney's Point** a point that lies approximately one-third the distance from the umbilicus to the right anterior superior iliac spine. It is thought to be the exact location of the appendix.

e. **Direct inguinal hernia** a hernia that protrudes "directly" into the inguinal canal medial to the inferior epigastric vessels.

f. **Indirect inguinal hernia** a hernia that protrudes into the inguinal canal lateral to the inferior epigastric vessels via the deep inguinal ring.

UNIT 5
Lower Limbs

A. Anterior Thigh

1. Preparation

Begin your preparatory work for the dissection of the lower limb by studying the bones of the lower extremity. Be sure to include the pelvis and foot and their bony landmarks in your preparation.

2. Cadaver Position

Supine with a block placed under the head.

3. Pre-dissection Discussion

When skinning and making cuts here, preserve the great saphenous vein (GSV), and as many of its tributaries as you can. Also, preserve the saphenous nerve running distally with it. They are both very superficial. The GSV is formed from the dorsal venous arch on the foot and drains into the femoral vein within the femoral triangle. The Saphenous nerve (L3, L4) is a sensory branch from the femoral. Also, there are many superficial veins (and many cutaneous nerves from obturator and femoral nerves) here, but the only superficial vein we are interested in is the great saphenous vein and the small saphenous vein on the posterior leg later in the dissection. There are also many lymph nodes here too. Ignore them.

4. Incisions to Make

Cut from the ASIS to the pubic tubercle, then cut from the midpoint of the inguinal ligament to the dorsum of the foot all the way to the base of the 3rd metatarsal. Then, make two or three transverse cuts across the longitudinal cut that you just made. Leave skin attached as best you can to use as cover to prevent drying out.

5. *Structures to Clean and Identify*

Muscles

Iliopsoas

Pectineus muscle

Quadriceps group
    rectus femoris
    vastus lateralis
    vastus medialis
    vastus intermedius

Sartorius

Adductor Magnus the hamstring portion is innervated by the tibial nerve while the adductor portion is innervated by the obturator nerve.

Adductor Longus

Adductor Brevis

Gracilis the most medial of the adductor muscles

Connective tissue

Fascia Lata

Iliotibial tract

Crural fascia

Inguinal ligament

Cribriform Fascia

Femoral Sheath it contains the femoral artery, femoral vein and the femoral canal which contains lymphatics.

Spaces

Saphenous Opening

Femoral Triangle it contains the femoral artery, femoral vein and femoral nerve.

Adductor Canal (Hunter's Canal or subsartorial canal) a muscular passageway beginning at the apex of the Femoral Triangle.

Contents:
a. femoral artery
b. femoral vein
c. nerve to vastus medialis this nerve is fairly large and heads directly into the vastus medialis,
d. saphenous nerve is usually found directly on the surface of the femoral artery as it traverses the adductor canal.

Note: These two nerves do not exit via the Adductor Hiatus.

Adductor Hiatus an opening in the adductor magnus between the adductor tubercle and the linea aspera. It is best seen posteromedially.

Nerves

Femoral (L2–L4)

Obturator (L2–L4)

Note: The anterior branch of the obturator nerve is found between the adductor longus and brevis, while the posterior branch is found between the adductor brevis and adductor magnus.

Saphenous (L3–L4) sensory branch from the femoral nerve leaves the apex of the subsartorial/adductor or Hunter's canal and runs with the great saphenous vein.

Blood Vessels

Dorsal venous arch of the foot

Great Saphenous vein (GSV)—beginning on the dorsum of the foot on the great toe side, then travels *anterior* to the medial malleolus and finally draining into femoral vein at the saphenous opening within the femoral triangle

Femoral artery (and its branches)

Femoral profundus artery
a. *Medial circumflex artery*
b. *Lateral circumflex artery*

6. *Clinical Applications*

a. **Inguinal Hernia** *(see previous discussion)*
   Direct found *medial* to inferior epigastric vessels (rare).
   Indirect found *lateral* to inferior epigastric vessels.

b. **Femoral Hernia** a protrusion within the femoral ring of abdominal viscera into the femoral canal. Most often occurring in women.

c. **Meralgia Paresthetica** compression of the lateral femoral cutaneous nerve resulting in paresthesia of the anterolateral thigh.

d. **Legg-Calve-Perthes Disease** a childhood hip disorder resulting from avascular necrosis of the proximal femur. The incidence is approximately 1 in 10,000.

e. **Patrick's test** the patient is supine and the examiner flexes, abducts and externally rotates the patient's femur. Pain during the test indicates hip dysfunction.

B. Anterior Leg—Lateral Leg—Dorsum of Foot

1. Preparation

Begin your preparatory work by studying the bones of the lower extremity. Be sure to include the pelvis and the foot.

2. Cadaver Position

Supine with a block placed under the head and one under the leg to raise it up for easier dissection.

3. Pre-dissection Discussion

The muscles in these compartments serve to dorsiflex the foot at the ankle as well as evert the foot at the ankle.

4. Incisions to Make

See above. Remove the skin, superficial fascia and the crural fascia of the lateral leg.

5. Structures to Clean and Identify (See Figure 23)

Muscles
Tibialis anterior
Extensor hallucis longus
Extensor hallucis brevis
Extensor digitorum longus

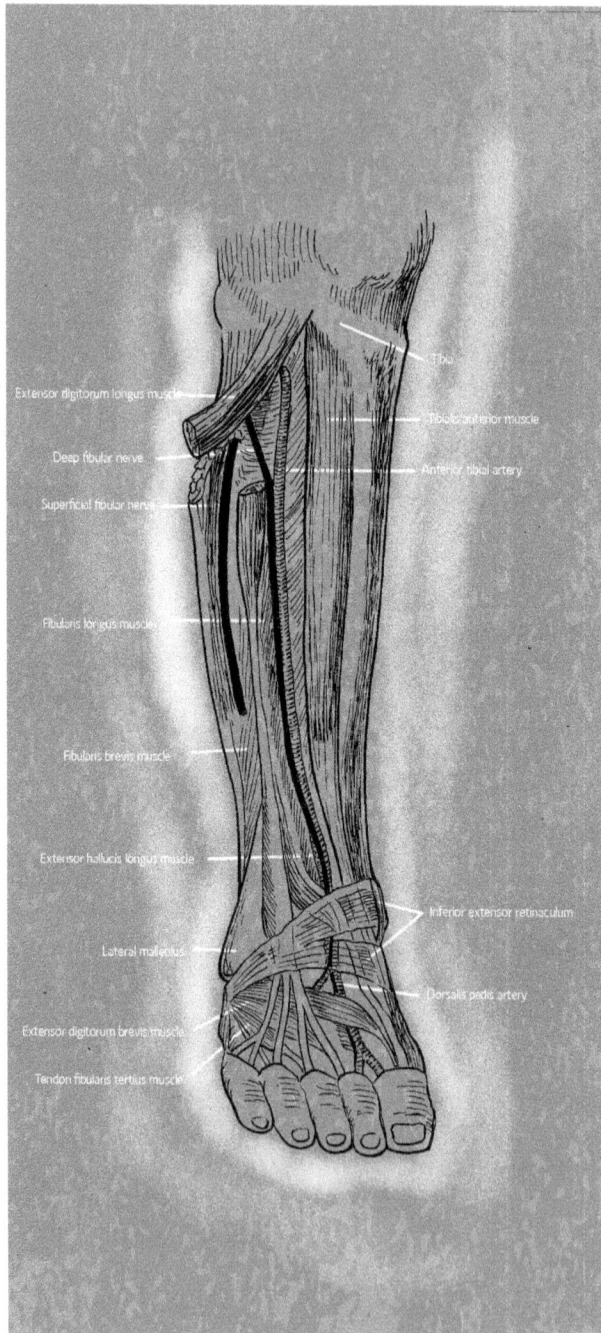

Figure 23. Anterior leg.

Extensor digitorum brevis

Fibularis longus

Fibularis brevis attached *laterally* to base of 5th metatarsal.

Fibularis tertius attached *dorsally* to the base of the 5th metatarsal and this tendon is *always anterior* to the lateral malleolus.

Connective tissue
Superior Retinaculum and the Y-shaped Inferior Retinaculum
Fibular Retinaculum

Nerves
Common fibular
Superficial Fibular found between the fibularis longus and fibularis brevis muscle bellies distally.

Deep Fibular found between extensor digitorum longus and tibialis anterior and it provides for a small sensory region between the great toe and 2nd toe.

Blood Vessels
Great saphenous vein begins on dorsum of foot-great toe side then passes *anterior* to the **medial** malleolus.

Anterior tibial artery

Dorsalis pedis artery found between the extensor hallucis longus and extensor digitorum longus tendons.

Arcuate (arch-like) artery found on dorsum of foot and provides digital branches to the toes.

6. *Clinical Applications*

a. **Foot Drop** a very common peripheral neuropathy often from compression of the common fibular nerve about the head of the fibula. Patients often exhibit the inability to dorsiflex the foot at the ankle.

b. **UnHappy Triad** a concurrent damage to the anterior cruciate ligament, medial meniscus and the medial collateral ligament often from a blow to the lateral aspect of the knee.

c. **Osgood-Schlatter Disease** a common cause of knee pain in adolescent boys. Pain and swelling occur on the tibial tuberosity due to repeated microtrauma usually from running or jumping types of activities.

d. **Chondromalacia Patellae** (*runner's knee*) knee pain of patellofemoral origin without history of trauma that seems to be localized to anterior to the patella, retropatellar or peripatellar. Prolonged sitting, squatting and running seem to be the precipitators of this syndrome. It is commonly assessed using Clarke's sign.

e. **Morton's Neuroma** compression and thickening of the interdigital nerve usually between the 3rd and 4th metatarsal due to ill-fitting shoes often seen in women.

f. **Drawer Test** the patient's knee is flexed to 90 degrees and the hip to 45 degrees. The examiner sits on the patient's foot whilst pulling the tibia forward and pushing it posteriorly, revealing no less than 6 mm movement in both directions. If more movement occurs, then the cruciate ligaments may be torn along with other associated ligaments of the knee.

g. **Clark's sign** a test is used to assess the presence of chondromalacia patellae. The patient lies relaxed with the leg extended while the examiner exerts a superior to inferior pressure on the patella as the patient contracts their quadriceps. If the test causes retropatellar pain, it is positive for chondromalacia patellae.

C. Gluteal Region

1. Preparation

Begin your preparatory work by studying the bones of the lower extremity. Be sure to include the pelvis and foot and their bony landmarks. Be especially aware of the structures that traverse the greater and lesser sciatic foramen. Prepare to clean and remove a lot of adipose tissue in the gluteal region. You will be uncovering the gluteal muscles, important ligaments and a small group of muscles that rotate the thigh laterally. These muscles can be remembered by using the mnemonic *P-GO-GO-Q* muscles.

2. Cadaver Position

Prone with a block placed under the thorax/sternum and head. Be certain to keep the face off of the dissection table.

3. Pre-dissection Discussion

In preparing to make cuts to open the gluteal region, be sure to be able to define the borders of the muscles and be certain to identify the location of the sacrotuberous ligament. It is a key structure in this dissection. Do not cut it. In the cadaver, the greater and lesser sciatic notches are now known as foramen, the greater and lesser sciatic foramen. They are formed by the sacrospinous ligament and sacrotuberous ligaments. Once you reflect the gluteus maximus muscle, you will see the piriformis muscle. Above this muscle is the suprapiriform recess and below it is the infrapiriform recess where the superior and inferior gluteal neurovascular bundles exit, respectively.

For Reference:

Greater Sciatic Foramen Contents   Lesser Sciatic Foramen Contents

a. Sciatic n.

b. Piriformis

c. Posterior cutaneous n. of thigh.

d. Superior/Inferior gluteal neurovascular bundle

e. Nerve to quadratus femoris

f. Pudendal nerve

g. Internal pudendal vessels

a. Internal pudendal vessels

b. Pudendal nerve

c. Nerve to Obturator internus

d. Obturator internus tendon

4. Incisions to Make

Remove the skin and deep fascia from the muscle. Again, locate the sacrotuberous ligament and preserve it. Once this has been done, identify the border of the gluteus maximus muscle, place your fingers along the fascial planes and separate it from the underlying gluteus medius muscle. At this point, you can now cut the gluteus maximus from its medial attachments to the ilium and sacrum. The gluteal vessels are deep to this muscle. Cut them as close to the muscle as possible. Notice the gluteus medius. Cut its attachment medially and flip it laterally. Clearly visible now is the gluteus minimus, as are the lateral rotator muscles of the thigh and the sciatic nerve. The gluteus minimus will not be reflected. Now, use blunt dissection to clearly separate the muscle bellies of the lateral rotators and clean the dissection field **(See Figure 21)**.

5. *Structures to Clean and Identify* (See Figures 24 and 25)

Muscles

Gluteus maximus
Gluteus medius
Gluteus minimus
Piriformis

Note: P-GO-GO-Q muscles

a. Piriformis—the *"key"* to other named structures in the area (S1–S2)

b. Gemellus Superior—inferior to piriformis (n. Obt. Intern)

c. Obturator Internus-inferior to Gemellus Superior (n. Obt. Intern)

d. Gemellus Inferior—inferior to Obturator Internus (n. Obt. Intern)

e. Obturator Externus—(not seen appreciably in this view-but its tendon can be found between the quadratus femoris and the inferior gemellus. (Obt. n)

f. Quadratus Femoris—inferior to the Inferior Gemellus (n. Quad. Fem)

Alcock's Canal—located medially to the ischial tuberosity. It contains the Pudendal nerve and Internal Pudendal artery.

Connective tissue

Gluteal aponeurosis

Tensor fascia lata

Sacrotuberous ligament

Sacrospinous ligament

Nerves

Sciatic nerve (L4-S3)

Posterior femoral cutaneous nerves (S1–S3) found running medial to sciatic nerve.

Inferior cluneal nerves from posterior femoral cutaneous nerve.

Superior gluteal nerve above the piriformis from Sacral Plexus.

Inferior gluteal nerve below the piriformis from Sacral Plexus.

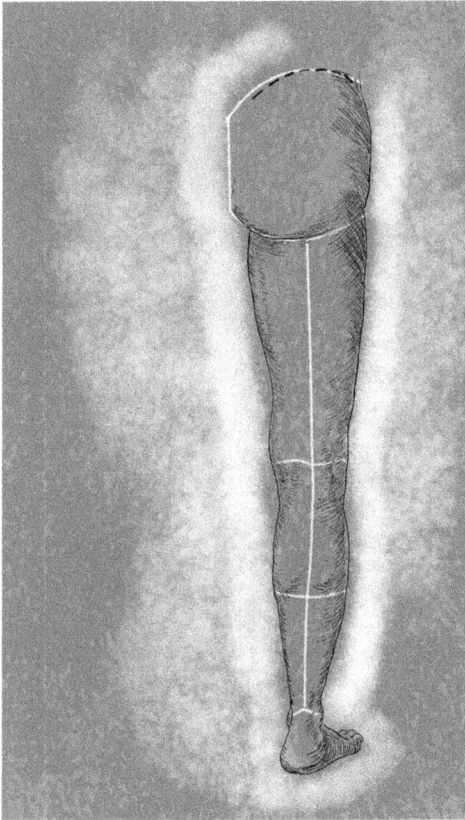

Figure 24. Incisions to make on the lower extremity posteriorly.

Pudendal nerve (S2–S4) passes through the greater AND lesser sciatic foramen into Alcock's Canal.

Blood Vessels

Superior gluteal artery

Inferior gluteal artery

Internal pudendal artery

Artery of the sciatic nerve can be from medial circumflex or inferior gluteal arteries.

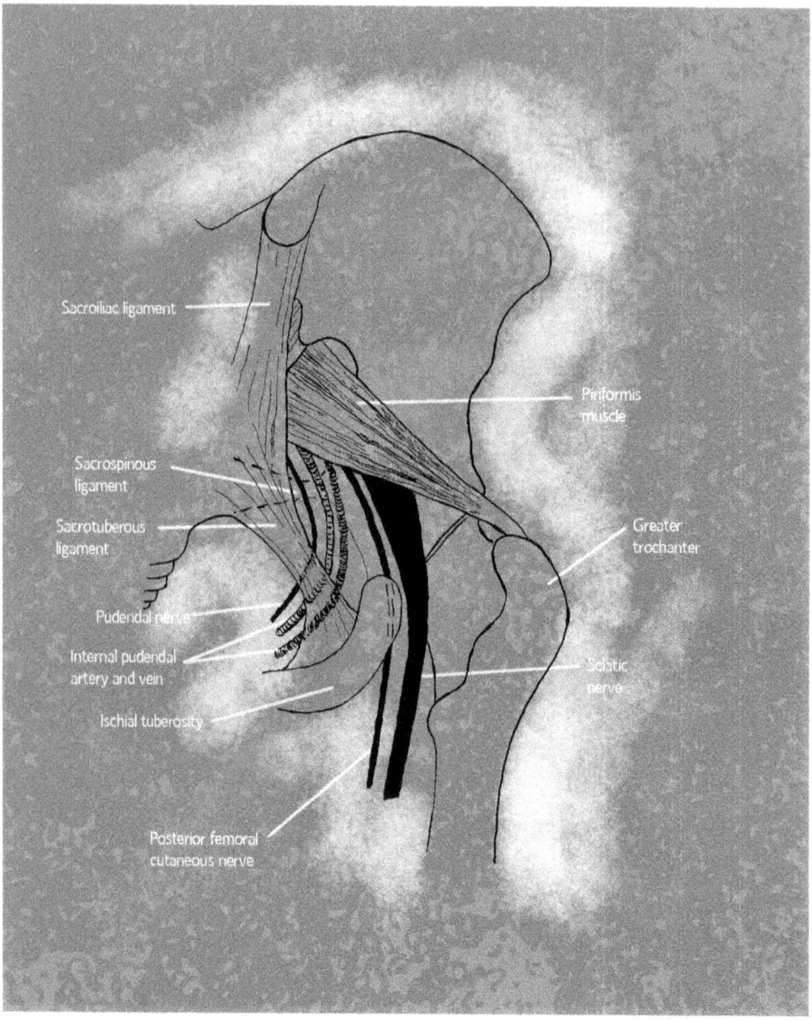

Figure 25. Deep posterior gluteal region. Gluteal muscles have been removed.

6. Clinical Applications

a. **Trendelenburg's Sign** the patient is standing and is instructed to raise the foot of the unaffected side off of the floor. If normal, the iliac crest may be low on the standing side and high on the side of the elevated leg. If the test is positive the iliac crest will be high on the standing side and low on the elevated leg side, indicating a coxal pathology.

b. **Piriformis Syndrome** an entrapment of the sciatic nerve along its trajectory in the gluteal region. The sciatic nerve usually passes under the piriformis but may occasionally travel through the muscle itself resulting in irritation of the nerve with subsequent pain in the gluteal and posterior thigh region.

c. **Pudendal Neuralgia** entrapment or compression of the pudendal nerve due to trauma, prolonged sitting (as on a bicycle seat) or horseback riding, resulting in impaired sexual function, and/or sphincter malfunction.

D. Posterior Thigh and Popliteal Fossa

1. *Preparation*

Begin your preparatory work by studying the bones of the lower extremity. Be sure to include the pelvis and the foot and their bony landmarks.

2. *Cadaver Position*

Prone with a block placed under the thorax and head.

3. *Pre-dissection Discussion*

The muscles located in the posterior compartment are usually referred to as the "hamstrings." There is not much difficulty separating the muscles out in this group. They are easily seen and easily identified. The short head of the biceps femoris is not considered a true hamstring muscle due to not attaching to the ischial tuberosity and being innervated by the common peroneal (fibular) nerve. In cleaning the popliteal artery, be sure to clean the vessels that form the geniculate anastomosis. In the posterior leg, be sure to locate the contributions to the sural nerve from both the common fibular and tibial nerves. They can be difficult to locate.

For Reference:

The geniculate anastomosis about the knee joint is formed by **eight** arteries:

1 superior medial genicular—from popliteal artery
1 inferior medial genicular—from popliteal artery
1 superior lateral genicular–from popliteal artery
1 inferior lateral genicular–from popliteal

1 descending genicular—from femoral artery
1 genicular—from lateral circumflex artery
1 anterior tibial recurrent— from anterior tibial artery
1 posterior tibial recurrent—from posterior tibial artery

4. Incisions to Make

In making the cuts, begin at the cluneal fold to the calcaneus. Now, at the level of the mid-posterior leg cut perpendicularly. Leave the skin attached as best you can so it can be used as cover. Be careful not to sever the short (lesser) saphenous vein and sural nerve(s) as both are quite superficial.

5. Structures to Clean and Identify (See Figure 26)

Muscles

Biceps Femoris
    long head
    short head (note the short head's innervation and origin)

Semitendinosus

Semimembranosus

Adductor Magnus (hamstring portion)

Popliteus muscle forms the floor of popliteal fossa and serves to medially rotate tibia, described as "unlocking the knee."

Pes Anserinus "Goose's Foot" the common insertion of the tendons of gracilis, sartorius and semitendinosus to the medial condyle of the tibia.

Spaces

Popliteal Fossa (and boundaries)
    Lateral—biceps femoris tendon
    Superomedial—semimembranosus and semitendinosus tendons
    Inferior—medial and lateral heads of the gastrocnemius
    Floor—popliteus muscle

Nerves

Sciatic nerve (L4–S3)

Tibial nerve this nerve usually provides a medial sural branch in the leg.

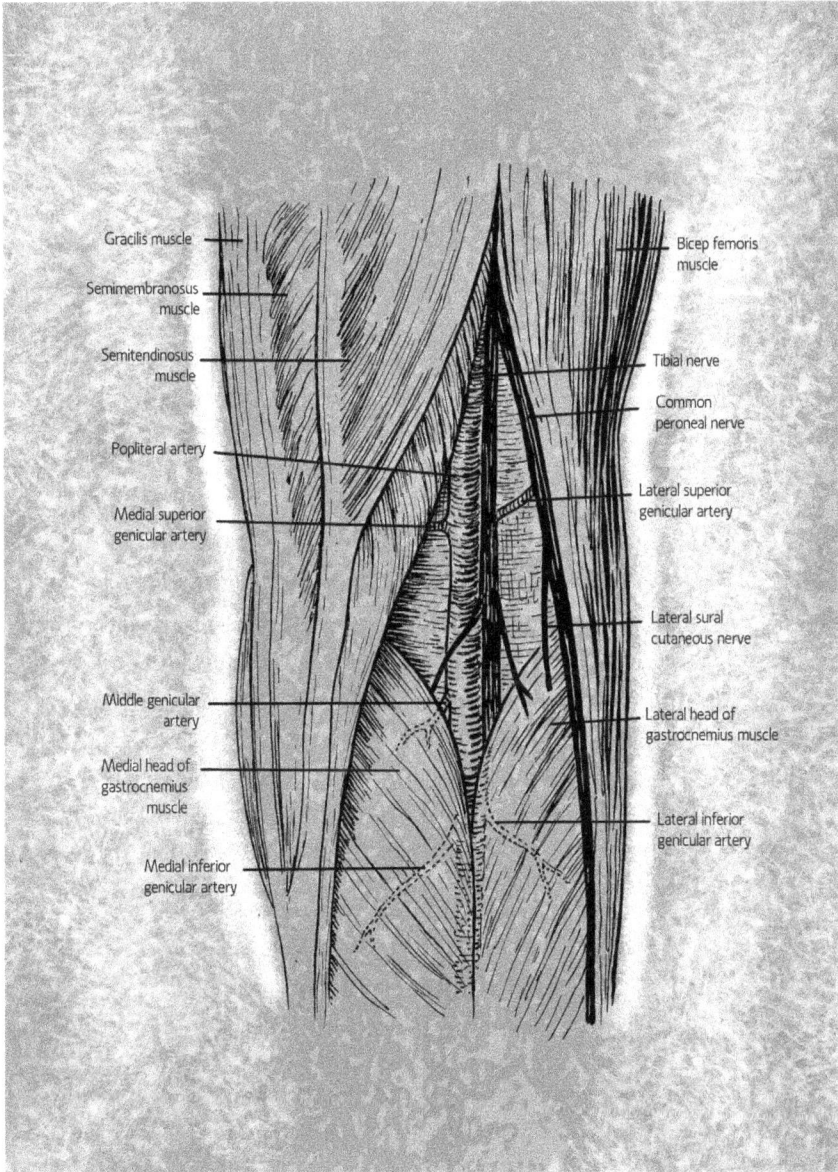

Figure 26. Popliteal fossa.

Common fibular this nerve usually provides a lateral sural branch in the leg.

Posterior femoral cutaneous (S1–S3) a fairly large nerve easily identified medial to the sciatic nerve in the infrapiriform recess.

Blood Vessels

Popliteal artery

  a. Superior and Inferior Medial Genicular

  b. Superior and Inferior Lateral Genicular

Femoral descending genicular

Lateral Femoral Circumflex genicular branch

Anterior Tibial anterior tibial recurrent

Posterior Tibial posterior tibial recurrent

Popliteal vein located superficial to the popliteal artery

Short Saphenous vein begins on the little toe side, running posterior to the lateral malleolus and travels with the sural nerve in the posterior leg to terminate in the popliteal vein.

6. Clinical Applications

  a. **Sciatica** irritation of the sciatic nerve along its trajectory by muscle spasm, discogenic origin, tumor or late pregnancy. Pain usually does not radiate to the foot, but stays in the posterior calf.

  b. **Popliteal Aneurysm** (Hunter's aneurysm) localized dilation of the popliteal artery within the popliteal fossa.

  c. **Baker's Cyst** a small fluid-filled sack within the popliteal fossa, often located between the semimembranosus and the medial belly of the gastrocnemius muscles. They may appear after trauma and may also be associated with various types of arthritis.

E. Posterior Leg

1. Preparation

Begin your preparatory work by studying the bones of the lower extremity. Be sure to include the pelvis and the foot and their bony landmarks in your preparation.

2. Cadaver Position

Prone with a block placed under the thorax and head keeping the face off of the table.

3. Pre-dissection Discussion

This area has superficial and deep compartments. These muscles are prime movers of plantarflexion at the ankle, flexing the toes, and a small amount of inversion. Also, the tendons and neurovascular structures of the deep posterior leg can be easily identified posterior to the medial malleolus by remembering Tom, **D**ick **A**nd a **V**ery **N**ervous **H**arry as they are identified from anterior to posterior relative to the medial malleolus.

4. Incisions to Make

This incision was made earlier during dissection of the posterior thigh. Clean the dissection field if necessary.

5. Structures to Clean and Identify (See Figure 27)

Muscles

Gastrocnemius

Soleus helps form the triceps surae with the 2 heads of the gastrocnemius.

Plantaris does not insert via the calcaneal tendon to the calcaneus. It has a separate attachment medial to the calcaneal tendon.

Tibialis Posterior deepest muscle of the posterior compartment, aka "Tom".

Flexor Hallucis Longus aka "Harry"

Flexor Digitorum Longus aka "Dick"

Note:

Tarsal Tunnel from anterior to posterior contains the tendons of tibialis posterior, flexor digitorum longus, posterior tibial artery, tibial nerve and the flexor hallucis longus.

Nerves

Tibial nerve

Sural nerve found between the two heads of the gastrocnemius.

Common Fibular notice it wrapping around the head of the fibula. It may also provide a sural branch that will join the sural branch of the tibial.

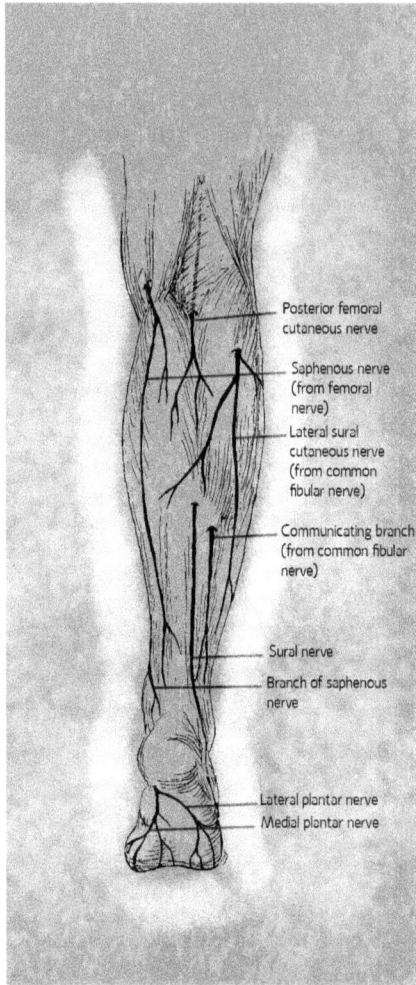

Figure 27. Superficial posterior leg.

Blood Vessels

Posterior tibial artery from popliteal and palpated immediately posterior to medial malleolus. It terminates as medial and lateral plantar arteries.

Fibular artery from the posterior tibial artery located between the tibialis posterior and the flexor hallucis longus.

F. Plantar Region

1. *Preparation*

Begin your preparatory work by studying the bones of the lower extremity. Be sure to include the tarsals, metatarsals, muscles, and ligaments.

2. *Cadaver Position*

Supine with a block placed under the thorax and head. You will also need to place a block under the calcaneus.

3. *Pre-dissection Discussion*

The plantar aponeurosis (deep fascia of the foot) is very tough. It is continuous with the crural fascia of the leg. The plantar fascia attaches to the calcaneus and divides into five digital bands attaching to each toe. Try to remove it without damaging the digital nerves, arteries, and muscles deep to it. All cuts will be made to ensure all structures remain attached to the calcaneus so that the layers of the foot can be reflected toward the calcaneus while preserving the neurovascular structures. When making incisions here, be aware that the skin is thick on the plantar region but thin on the toes themselves.

For Reference:

The Four Layers of the Foot (superficial to deep):

1. Abductor Hallucis, Flexor Digitorum Brevis, Abductor Digiti Minimi.
2. Quadratus Plantae, four lumbricals, the tendon sheaths of the FHL and FDL.
3. Flexor Hallucis Brevis, Adductor Hallucis, Flexor Digiti Minimi, Long Plantar Ligament, Short Plantar Ligament.
4. (4) Dorsal and (3) Plantar Interossei, tendons of Tibialis Posterior and Peroneus Longus.

Note: The muscles of the sole of the foot are in 4 layers. The names of the muscles suggest that each individual toe can be controlled alone, that is very unlikely though.

4. Incisions to Make

In opening this area, cut from the head of the second metatarsal to the calcaneus. Then make a cut across the webbed area on the plantar surface, as well as smaller cuts longitudinally on each toe. Cut and reflect the plantar aponeurosis, cut the tendons (as far distally as you can) of the flexor digitorum brevis and reflect it, cut the tendons of the flexor digitorum longus and reflect it. You now can see the muscles within the deeper layers. These muscles are not to be cut, only separated and identified **(See Figure 24)**.

5. Structures to Clean and Identify (See Figures 28 and 29)

Muscles

Flexor Digitorum Brevis

Abductor Hallucis

Abductor Digiti Minimi

Plantar Fascia (and its 5 digital bands for each toe)
Lumbricals (originating from the tendons of the FDL)
Flexor Hallucis Brevis (maybe find some sesamoid bones here?!)

Adductor Hallucis (transverse and oblique head)

Flexor Digiti Minimi Brevis

Interossei (Plantar and Dorsal— PAD, DAB)

Quadratus Plantae (Flexor Accessorius) this muscle inserts into the tendons of flexor digitorum longus and seems to allow flexing of the four toes in any position of the ankle and helps to change the direction of action of the FDL too.

Tendons of Flexor Digitorum Longus its four tendons pass through the flexor digitorum brevis.

Tendon of Tibialis Posterior

Tendon of Fibularis Longus travels all the way to the base of the first metatarsal and the medial cuneiform.

Connective tissue

Plantar Calcaneonavicular "Spring" Ligament from the sustentaculum tali to inferior surface of the navicular.

Short Plantar Ligament from plantar surface of calcaneus to the cuboid.

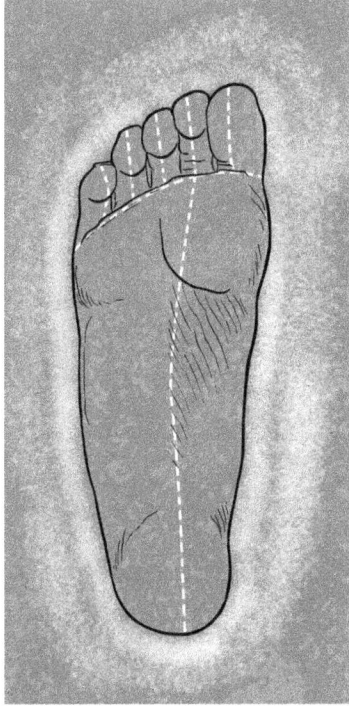

Figure 28. Incisions to make on plantar foot.

Long Plantar Ligament from plantar surface of calcaneus to bases of 2nd, 3rd and 4th metatarsals.

Deltoid Ligament (four parts)

1. Anterior tibiotalar
2. Tibiocalcaneal
3. Posterior tibiotalar
4. Tibionavicular

This four-part ligament resists "over-eversion" of the hindfoot (the talus and calcaneus)

Nerves

Medial Plantar located between the abductor hallucis and flexor digitorum brevis and is larger than the Lateral Plantar nerve.

Proper Plantar medial and lateral both from the Medial Plantar nerve.

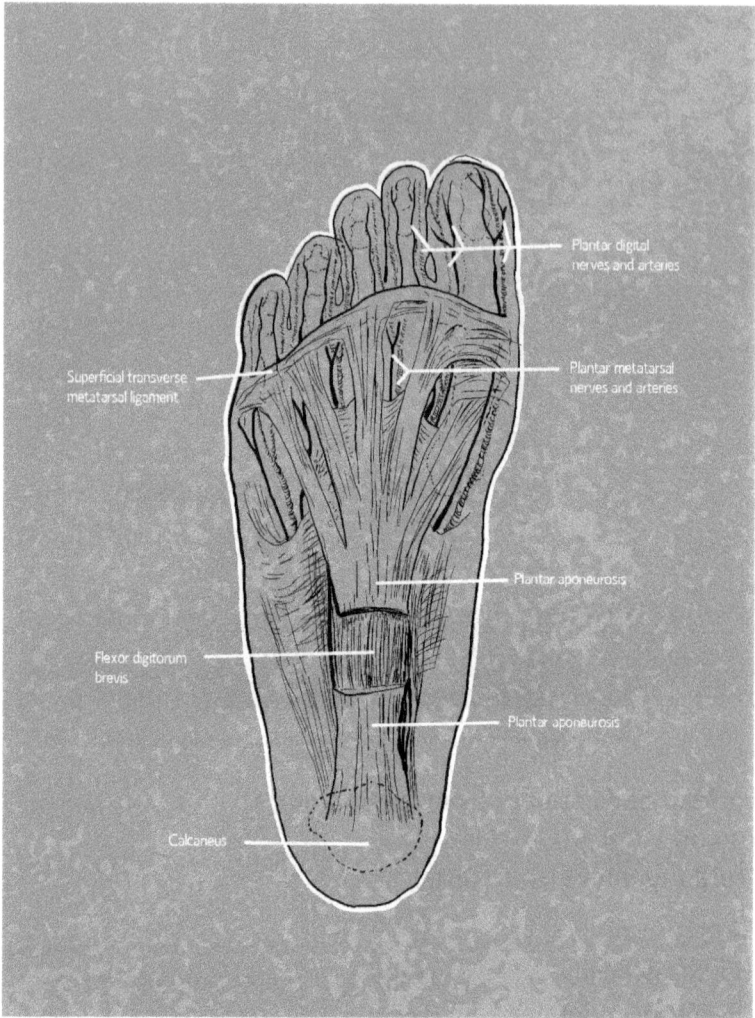

Figure 29. Superficial plantar foot.

Lateral Plantar projects laterally across the plantar surface between the quadratus plantae and flexor digitorum brevis.

Note: The lateral plantar nerve innervates most intrinsic muscles of the foot except flexor hallucis brevis, flexor digitorum brevis, abductor hallucis, and 1st Lumbrical.

Blood Vessels

Lateral Plantar artery It is larger than the medial. It heads laterally and then medially to form the plantar arch, giving proper digital branches.

Medial Plantar artery found between the adductor hallucis and the flexor digitorum brevis.

6. *Clinical Applications*

a. **Plantar Fasciitis** inflammation of the plantar aponeurosis from mechanical repetitive stress, degeneration or ill-fitting shoes. Pain is especially apparent in the morning upon rising from bed.

b. **Tarsal Tunnel Syndrome** entrapment of the posterior tibial neurovascular bundle within the tarsal tunnel resulting in numbness and tingling in the foot and toes.

c. **Hallux Valgus** a deviation of the metatarsal head of the great toe resulting in a lateral deviation of the proximal phalanx of the same toe causing a bunion. It is most often seen in women who wear ill-fitting shoes, especially high-heeled shoes.

d. **Hammer Toe** hyperflexion of the proximal interphalangeal joint, with associated distal interphalangeal hyperextension. Mostly from wearing ill-fitted shoes, trauma or arthritis.

e. **Pes Planus** flat feet due to the disruption of the medial longitudinal arch of the foot.

f. **Pes Cavus** a foot deformity due to a high longitudinal arch of the foot.

g. **Morton's Test** the examiner exerts transverse pressure across the heads of the metatarsals. If sharp pain is elicited during the test it is positive for metatarsalgia or a neuroma.

UNIT 6
Neck

A. Anterior Neck and External Larynx

1. Preparation

Begin your preparatory work by studying the cervical vertebrae, muscles of the anterolateral neck, the triangles of the neck, their borders and contents, the larynx and its cartilages, the trachea, and various nerves in the area.

2. Cadaver Position

Supine with a block placed under the thorax and head.

3. Pre-dissection Discussion

This area, both anterior and lateral, has many lymph nodes, and they should be removed if they are crowding the dissection field. The muscles in the anterior region are often grouped into two groups, superficial and deep. The muscles in this region also present various triangles. The contents of the triangles will be cleaned and studied. In most cases, structures will remain *in situ*.

For Reference:

The Superficial Group is divided into two groups (innervations are noted)

Suprahyoid Group—elevating the hyoid bone

1. Digastric—two bellies (Trigeminal and Facial)
2. Stylohyoid—Facial
3. Geniohyoid—C1
4. Mylohyoid-Trigeminal

Infrahyoid Group—depressing the hyoid bone

1. Sternohyoid—C1
2. Sternothyroid—C1
3. Thyrohyoid—C1
4. Omohyoid—two bellies C1–C3

The Infrahyoid group is innervated by the Ansa Cervicalis (C1–C3), while the Suprahyoid group typically has named cranial nerve branches innervating them, mostly.

The Deep group will be mentioned in the Special Dissection Unit.

1. Longus colli
2. Longus capitis
3. Rectus capitis anterior
4. Rectus capitis lateralis

Cervical Triangles

The sternocleidomastoid muscle divides the neck into anterior and posterior triangles. The posterior triangle was dissected in Unit 1. The anterior triangle is of most interest here as it has several other triangles within it. The anterior triangle is bound by the midline of the neck, inferior border of the mandible, and the anterior border of the sternocleidomastoid muscle. The smaller triangles within the anterior triangle include:

1. Submental
2. Carotid
3. Muscular
4. Submandibular

♦ Be sure to review their borders and contents in an atlas

4. Incisions to Make

Make an incision at the apex of the mastoid process to the mental protuberance, then from the mental protuberance to the jugular notch. Finally, cut from the acromion process to the sternal notch. Do not go too deeply! The skin thickness varies greatly in this area. Ultimately, the goal is to remove the skin and deep fascia. No skin is to be kept. Watch for the platysma and the cervical branch of the facial nerve passing deep to it. Preserve the external jugular vein and the transverse cervical nerve crossing the sternocleidomastoid muscle in the deep fascia, at Erb's point. The Carotid Sheath may be readily seen and can be opened, and the contents studied now or later.

5. *Structures to Clean and Identify* (See Figures 30 and 30.1 and 31)

Muscles, glands and connective tissue, and cartilages

Platysma

Hyoid bone

Thyrohyoid Membrane

Cricothyroid membrane

Thyroid cartilage

Laryngeal prominence

Thyrohyoid membrane provides for a hole thyrohyoid foramen, for the passage of the internal laryngeal nerve and the superior laryngeal artery into the larynx.

Median Thyrohyoid ligament

Lateral Thyrohyoid ligament

Cricoid cartilage

Epiglottic cartilage

Tracheal rings and carina

Cricothyroid muscle

Thyroid Gland (lobes, isthmus and maybe a pyramidal lobe)

Parathyroid glands

Digastric presents an anterior and posterior belly with different innervations.

Mylohyoid forms the floor of the mouth.

Sternocleidomastoid this muscle divides neck into anterior and posterior triangles.

Scalene Muscles

1. Anterior—locate the phrenic nerve on its anterior
2. Middle—the dorsal scapular nerve pierces this muscle
3. Posterior—smallest

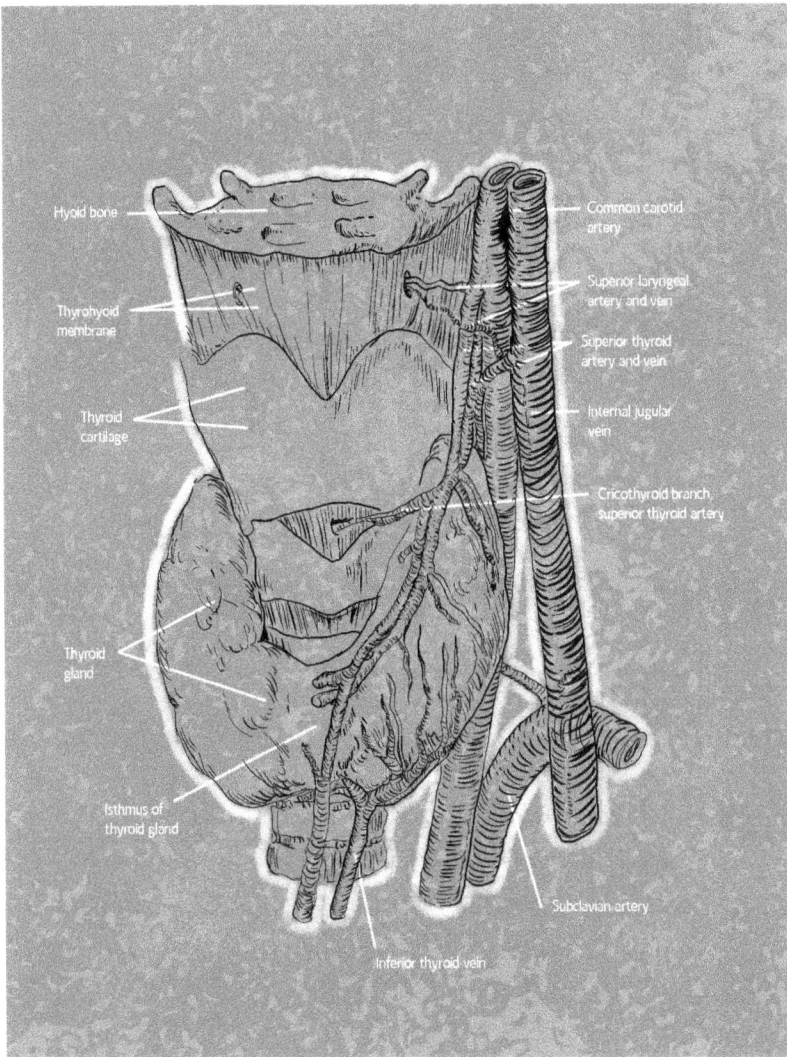

Figure 30. Thyroid cartilage and thyroid gland.

Sternohyoid

Omohyoid presents a superior and inferior belly.

Sternothyroid

Thyrohyoid

Carotid Sheath (and contents) formed from the fusion of the investing, pretracheal and prevertebral fascia. It extends from the base of the skull to the arch of the aorta.

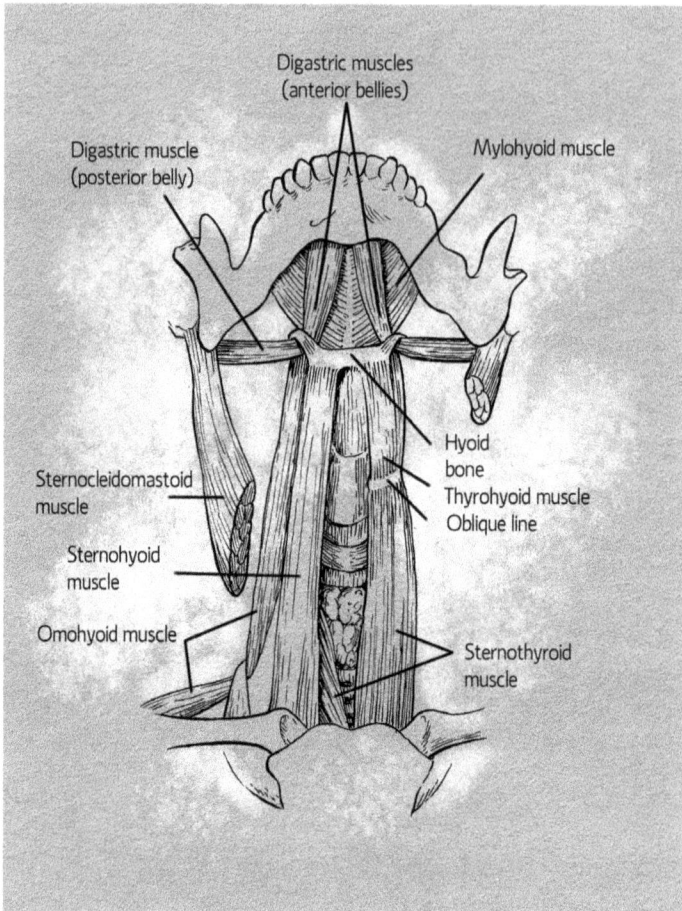

Figure 30.1. Deep anterior neck.

Contents of the carotid sheath:
Common carotid artery
Internal carotid artery (with sympathetic fibers clinging to it)
Internal jugular vein
Vagus nerve

Submandibular gland

Nerves

Ansa "loop" Cervicalis (C1–C3) it surrounds the carotid sheath (or within its wall)
Superior root (C1)—runs with hypoglossal nerve briefly
Inferior root (C2–C3)—will join with superior root

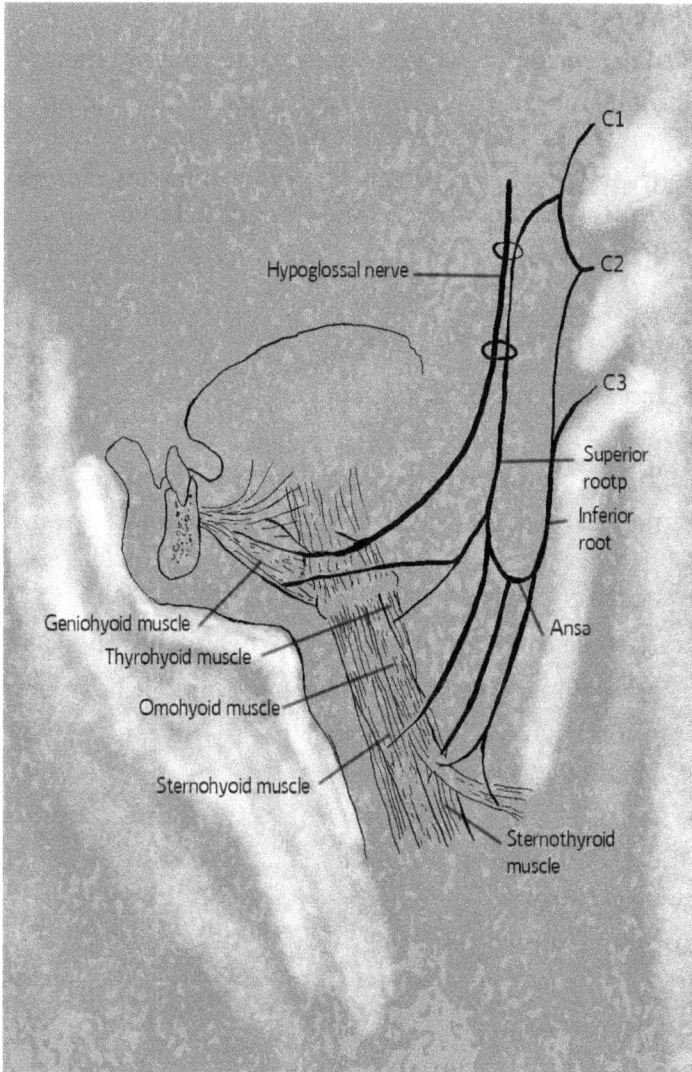

Figure 31. Deep lateral neck.

Cervical branch of Facial nerve it emerges from the lower border of the parotid gland to the just below the angle of the mandible and *enters the deep surface* of the platysma.

Supraclavicular (C3–C4) found within the superficial fascia of the platysma and identified as medial, intermediate, and lateral.

Superior laryngeal

a. Internal—sensory to larynx—pierces thyrohyoid membrane.

b. External—motor to cricothyroid muscle and small part of inferior pharyngeal constrictor muscle.

Recurrent laryngeal it enters the larynx with the inferior laryngeal artery to innervate the intrinsic muscles of the larynx.

Glossopharyngeal

Phrenic

Blood Vessels

Subclavian artery (three parts)

1st Part

a. Vertebral—up through the transverse foramen C6–C1

b. Internal thoracic—on either side of the internal surface of sternum

c. Thyrocervical trunk

　　1. transverse cervical—on posterior surface of trapezius

　　2. suprascapular—over the transverse scapular ligament

　　3. inferior thyroid—heads medially to thyroid gland (gives ascending cervical)

2nd Part

a. Costocervical trunk

　　1. deep cervical

　　2. supreme intercostal

3rd Part

　　1. dorsal scapular

Subclavian vein

Vertebral artery

External jugular vein found on the surface of the sternocleidomastoid muscle.

Superior thyroid artery

Superior laryngeal artery from superior thyroid artery and enters the larynx via thyrohyoid membrane with the internal laryngeal nerve and is a branch of the superior laryngeal nerve.

B. Lateral Neck

1. Preparation

Begin your preparatory work by studying the cervical vertebrae, muscles of the neck and the triangles of the neck. The SCM creates the boundary between the Anterior and Posterior triangles.

2. Cadaver Position

Supine with a block placed under the thorax and head. You may have to rotate the head for better positioning.

3. Pre-dissection Discussion

As you skin and dissect this area be careful to not destroy the cutaneous nerves of the nerve point of the neck. You will also find many lymph nodes here. Remove them from the dissection field. The only muscles that need detached are the SCM and platysma. Continue to clean the area with a blunt probe and your fingers after these muscles have been detached. The Carotid Sheath can be readily seen and can be opened and studied now if it hasn't been already.

For Reference:

Punctum Nervosum Point in the Neck found about at the posterior border of the SCM just a few centimeters above the clavicle. Several sensory nerves of the cervical plexus emerge here.

A. Great Auricular (C2, C3)

B. Lesser Occipital (C2, C3)

C. Transverse Cervical (C2, C3)

D. Supraclavicular (C3, C4)

4. Incisions to Make

Locate the sternocleidomastoid and platysma. Detach both from their lower attachments. Clear away the deep cervical fascia to clean the dissection field.

5. Structures to Clean and Identify

Muscles, glands, connective tissue and spaces

Scalenes

1. Anterior—locate the phrenic nerve on its anterior
2. Middle—(largest)—dorsal scapular nerve pierces this muscle
3. Posterior—very small and insignificant

Sternohyoid

Omohyoid

Sternothyroid

Thyrohyoid

Carotid Sheath

Submandibular gland

Nerves

Lesser occipital

Vagus

Recurrent laryngeal

Superior laryngeal

— **External laryngeal** from superior laryngeal and it innervates the cricothyroid muscle and small part of inferior pharyngeal constrictor muscles.

— **Internal laryngeal** from the superior laryngeal nerve, traveling through the thyrohyoid membrane providing sensation to the larynx above the vocal folds.

Ansa Cervicalis a loop of nerves (C1–C3) around the internal jugular vein within the wall of the carotid sheath.

Transverse Cervical nerve passes transversely across the SCM.

Great Auricular just superior to the transverse cervical nerve, travels parallel to the external jugular vein.

Dorsal Scapular nerve (C5 root) pierces the middle scalene and then travels parallel to the medial border of the scapula.

Phrenic (C3–C5) found on the surface of the anterior scalene— (C3, C4, C5— keeps the diaphragm alive).

Blood Vessels

Internal Jugular vein

Common Carotid artery divides into the internal and external carotid at about C4 vertebral level.

External Carotid artery found in the anterior triangle, ascending beneath CN XII and the posterior belly of digastric. It ends within the substance of the Parotid gland by dividing into its two terminal branches, the Maxillary and the Superficial Temporal.

Note: Branches of the External Carotid Artery from inferior to superior:

a. superior thyroid

b. lingual

c. facial

d. occipital

e. posterior auricular

f. ascending pharyngeal

g. superficial temporal

h. maxillary

Superior Thyroid artery

Superior laryngeal artery enters larynx via thyrohyoid foramen with the internal laryngeal nerve.

UNIT 7
Head

A. Anterior Face

1. Preparation

Begin your preparatory work by studying the facial skeleton. Be sure to include as many of the bones, muscles, and blood vessels as you can.

2. Cadaver Position

Supine with a block placed under the thorax and head.

3. Pre-dissection Discussion

The anterior face presents mostly muscles. These muscles are often termed "muscles of facial expression." It is important to note that these muscles move the skin and rarely move a bone. They do serve to express emotion as well as act as sphincter muscles to regulate the size of an opening too. Most of these muscles are innervated by branches of the facial nerve. The primary blood supply to the face is via the facial artery, a branch of the external carotid. Preserve the facial artery and vein as well as the terminal branches of the facial nerve. The objective is to remove the skin of the face and the subcutaneous tissue while preserving the muscles and neurovascular structures for study. The neural structures include branches of the facial and the trigeminal nerves.

4. Incisions to Make

Cut from the vertex to the nasion, then around each orbit, around the mouth, about the nose through the philtrum and from the lower lip to the mental protuberance. Do not cut too deeply. Remove the skin completely. After the skin is removed blunt dissection using the probe and hemostats is the preferred method of dissection.

5. *Structures to clean and Identify* (See Figures 32 and Figure 33 and Figure 34)

Muscles

Orbicularis oris

a. orbital portion—surrounds orbit

b. palpebral portion—in eyelid

c. lacrimal portion—may help with lacrimation by compressing the lacrimal sac

Frontalis

Occipitalis

Corrugator supercilii

Procerus

Depressor septi

Nasalis

a. transverse part

b. alar part

Buccinator

Levator labii superioris

Levator labii superioris alaeque nasi

Levator anguli oris

Zygomaticus major

Zygomaticus minor

Risorius

Depressor anguli oris

Depressor labii inferioris

Mentalis

Platysma

Levator palpebrae superioris

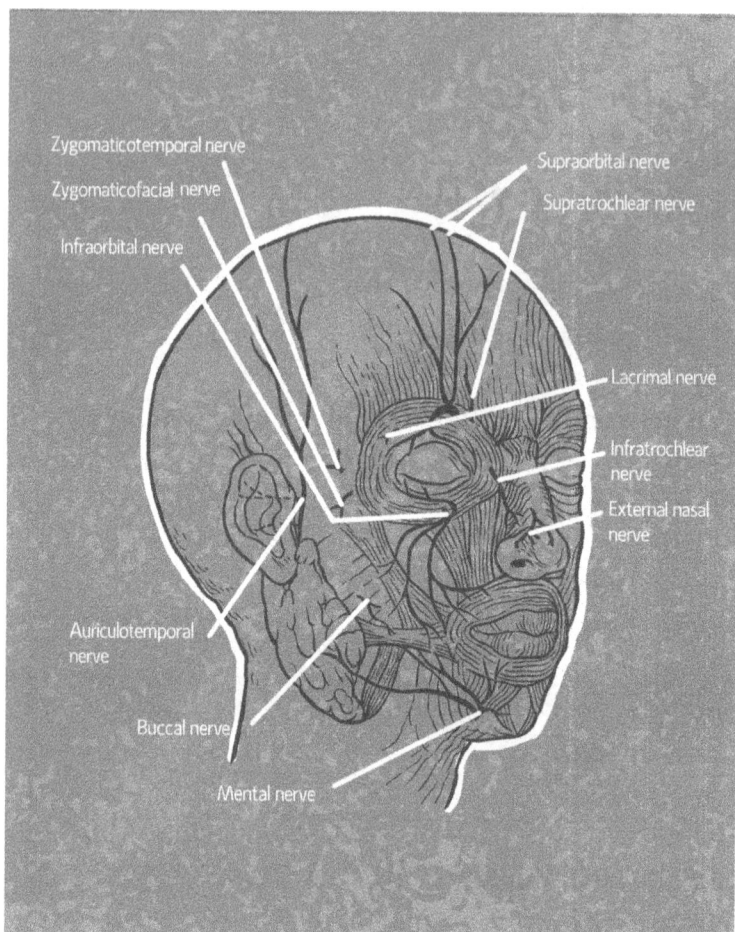

Figure 32. Superficial nerves of the face.

Nerves

Facial nerve and its terminal branches

a. temporal

b. zygomatic

c. buccal

d. marginal mandibular

e. cervical

f. posterior auricular

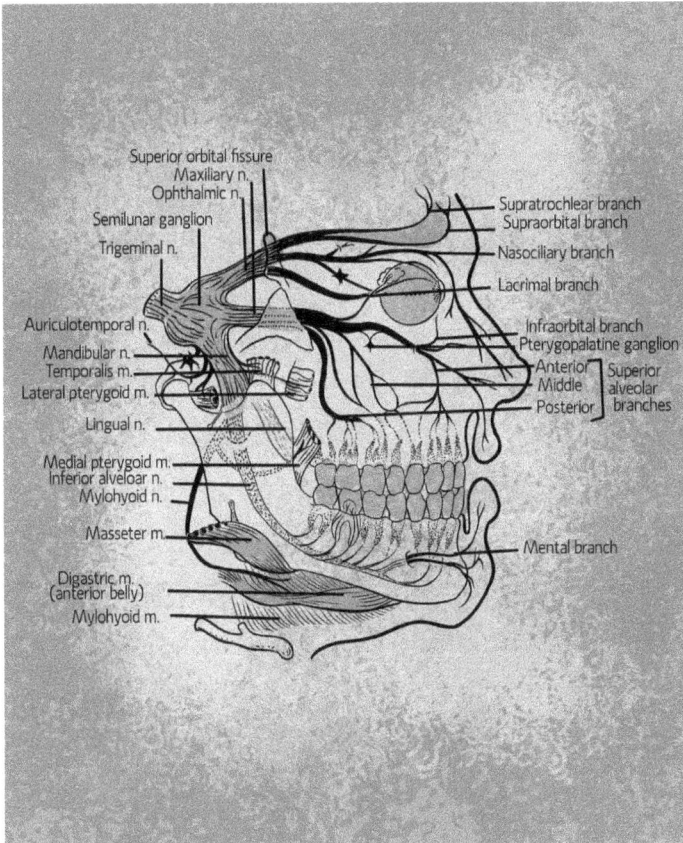

Figure 33. Branches of trigeminal nerve.

Trigeminal nerve and its facial/cranial branches

a. supraorbital

b. infraorbital

c. supratrochlear

d. zygomaticofacial

e. zygomaticotemporal

f. external nasal

g. buccal

h. auriculotemporal

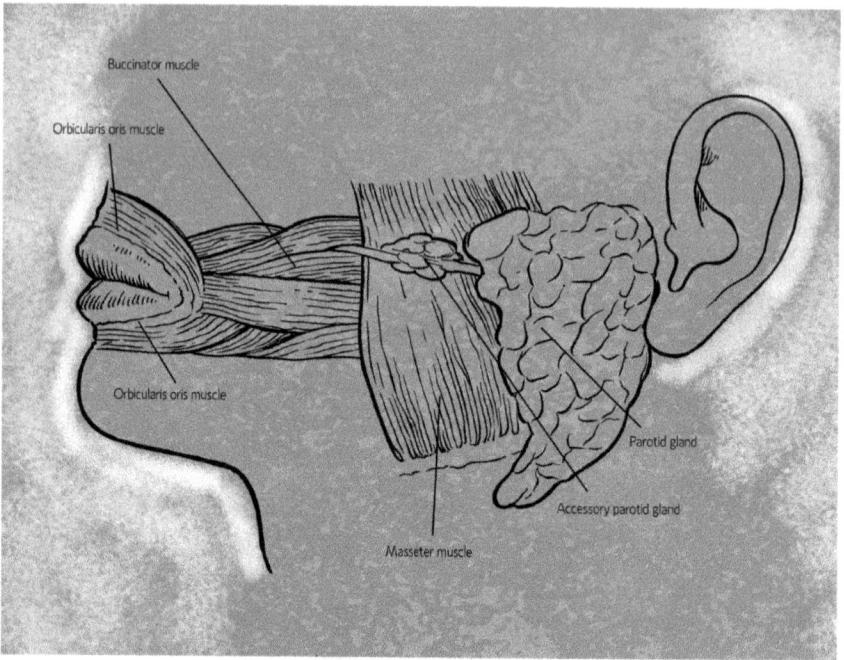

Figure 34. Parotid region.

Blood Vessels
Facial artery branches

a. superior labial

b. inferior labial

c. angular

d. lateral nasal

e. mental

f. dorsal nasal

g. tonsillar

h. submental

6. Clinical Applications

a. **Bell's Palsy** an acute lower motor neuron facial paralysis thought to be related to a viral infection trapping the "swollen" nerve in the facial canal. It must be differentiated from a stroke.

b. **Tic Douloureux** sudden, severe unilateral facial pain involving one or more branches of the trigeminal nerve.

c. **Kisselbach's area** an anastomotic network of arteries about the nose and nasal septum accounting for over 90 percent of nosebleeds.

B. Parotid Region

1. Preparation

Begin your preparatory work by studying the facial skeleton. Be sure to include all bones, muscles, blood vessels and the infratemporal fossa especially.

2. Cadaver Position

Supine with a block placed under the thorax and head. You may need to rotate the head to better dissect this region.

3. Pre-dissection Discussion

The Parotid Region is one of the most difficult and time-consuming areas to dissect. It is the area located on the lateral side of the face just anterior to the ear. It is difficult to dissect simply because the parotid gland is challenging to remove without damaging the structures that traverse it. It is enclosed within the Parotid Sheath, a very tough connective tissue that is not easily removed. When it is removed carefully you will see the structures of interest. These include the external carotid artery and its branches such as superficial temporal, posterior auricular and maxillary, as well as the retromandibular vein. You will also see the terminal branches of the facial nerve as it emerges from the stylomastoid foramen and finally, you will see the auriculotemporal nerve running with the superficial temporal artery. Also of note is the posterior division of the retromandibular vein and posterior auricular vein form the External Jugular Vein, forming just inferior to the parotid gland.

4. Incisions to Make

After carefully skinning the area, begin by removing the parotid sheath and picking apart the parotid gland, piece by piece. Be careful, the branches of the facial nerve are very superficial. An alternative (if you are not interested in preserving the terminal branches of the facial nerve) is to remove the parotid gland entirely from the parotid bed reflecting it superiorly but keeping it attached to Stenson's Duct.

5. *Structures to Clean and Identify* (See Figures 34 and 35)

Muscles and glands

Parotid Gland a purely serous gland and the largest of the salivary glands.

Parotid duct arises from the anterior part of the gland, runs over the masseter just below the zygomatic arch, pierces the buccinator and opens opposite the upper 2nd molar in the oral cavity.

Masseter

Posterior belly of digastric

Stylohyoid The stylohyoid muscle belly splits to allow the digastric muscle to pass through.

Nerves

Facial nerve and its terminal branches
    Temporal
    Zygomatic
    Buccal
    Marginal Mandibular
    Cervical
Memory trick –"Ten Zebras Bit My Mother's Chickens"

Auriculotemporal found between the head of the mandible and EAM running with the Superficial Temporal Artery. This nerve presents two roots, a superior root (sensory) and an inferior root (parasympathetic).

Blood Vessels

Retromandibular vein formed from the union of the maxillary vein and the superficial temporal vein.

Posterior auricular artery runs between posterior belly of digastric and parotid gland.

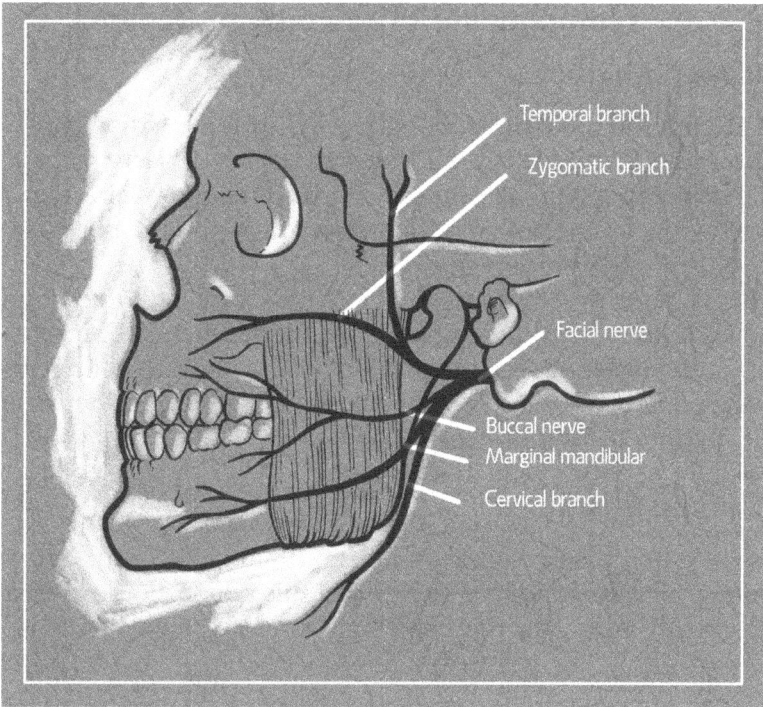

Figure 35. Terminal branches of facial nerve.

Superficial temporal artery one of the two terminal branches of the external carotid artery, the other being the maxillary artery.

External carotid artery branches seen here

a. **Superficial temporal** one of the two terminal branches of the external carotid artery.

b. **Posterior auricular** passes directly behind the ear.

c. **Maxillary** many branches in the infratemporal fossa.

d. **Occipital** passing up behind mastoid process.

6. *Clinical Applications*

a. **Frey's Syndrome** *(auriculotemporal syndrome)* often a sequela to parotid salivary gland surgery whereby the aberrant reinnervation of postganglionic parasympathetic neurons to nearby denervated sweat glands and blood vessels. This reinnervation causes flushing, sweating and burning in response to mastication and salivation.

C. Temporal and Infratemporal Regions

1. Preparation

The Temporal area must be dissected prior to dissecting the Infratemporal area. It is a shorter and easier dissection than the Infratemporal region dissection. The infratemporal crest on the greater wing of the sphenoid bone separates the temporal region from the infratemporal region.

Begin your preparatory work by studying the temporal bone, sphenoid bone and mandible. Be sure to include muscles of mastication, internal maxillary artery and its branches, trigeminal nerve, and parotid gland in your study. Also review the boundaries of the temporal and infratemporal fossae.

2. Cadaver Position

Supine with a block placed under the thorax and head. You may need to rotate the head to better dissect this region. Some dissectors prefer to disarticulate the head for this dissection, but it is not necessary.

3. Pre-dissection Discussion

The Temporal region contains fascia (both superficial and deep), several muscles, veins, and terminal branches of the facial nerve. The Infratemporal region contains a venous plexus, neurovascular bundles, muscles of mastication and the maxillary artery and its branches. The zygomatic arch will need to be removed at some point during this dissection, as will the ramus of the mandible. The skin and the superficial fascia will be removed but do not remove the superficial temporal fascia (temporoparietal fascia) lying deep to the subcutaneous tissue just yet. You will immediately see the superficial temporal vessels and the auriculotemporal nerve (the nerve lies posterior to the vessels) on the surface of the temporoparietal fascia. Also, on this fascia you will see the terminal branches of the facial nerve. You may preserve them. Identify and preserve the superficial temporal vein receiving the transverse facial vein. Locate the facial vein, found on the surface of the masseter and preserve it. You also will see the posterior branch of the retromandibular vein joining with the posterior auricular vein forming the external jugular vein, preserve this arrangement. After identifying and preserving the superficial structures, reflect the temporoparietal fascia superiorly. Next, reflect the masseter muscle inferiorly from the ramus of the mandible and zygomatic arch but keep it attached to the inferior border of the mandible being careful to preserve the masseteric artery and nerve passing through the mandibular notch. Cut the zygomatic arch, remove the deep fascia covering the temporalis muscle and reflect the temporalis superiorly cutting its attachment to the coronoid process of the mandible while preserving the Buccal branch of the Trigeminal which is usually within the tendon of the temporalis muscle. Notice the deep temporal artery and nerve on its posterior surface. Clean the dissection field completely to visualize the neurovascular structures. You will have to remove the pterygoid muscles piecemeal to accomplish this.

4. Incisions to Make

Cut the zygomatic arch by placing two probes under the zygomatic arch, one as far anteriorly as possible and another placed as far posteriorly as possible. Cut between the probes and remove the arch. Cut the neck of the mandible, remove the piece, then slide a probe behind the ramus of the mandible and slide as far inferiorly as the lingula. Cut just above the probe but do not cut completely through. Doing this will preserve the Inferior Alveolar neurovascular bundle traversing the mandibular canal. Or, alternatively, for a more cautious dissection, open the mandibular canal with a chisel, not a saw. Begin at the mental foramen, moving superiorly, removing only the outer table of bone to expose the canal and the inferior neurovascular bundle.

5. Structures to Clean and Identify (See Figure 36)

Muscles, glands and connective tissue
Temporoparietal fascia
Temporoparietalis
Temporalis muscle
Deep temporal fascia
Medial (internal) Pterygoid muscle
  a. Deep head
  b. Superficial head

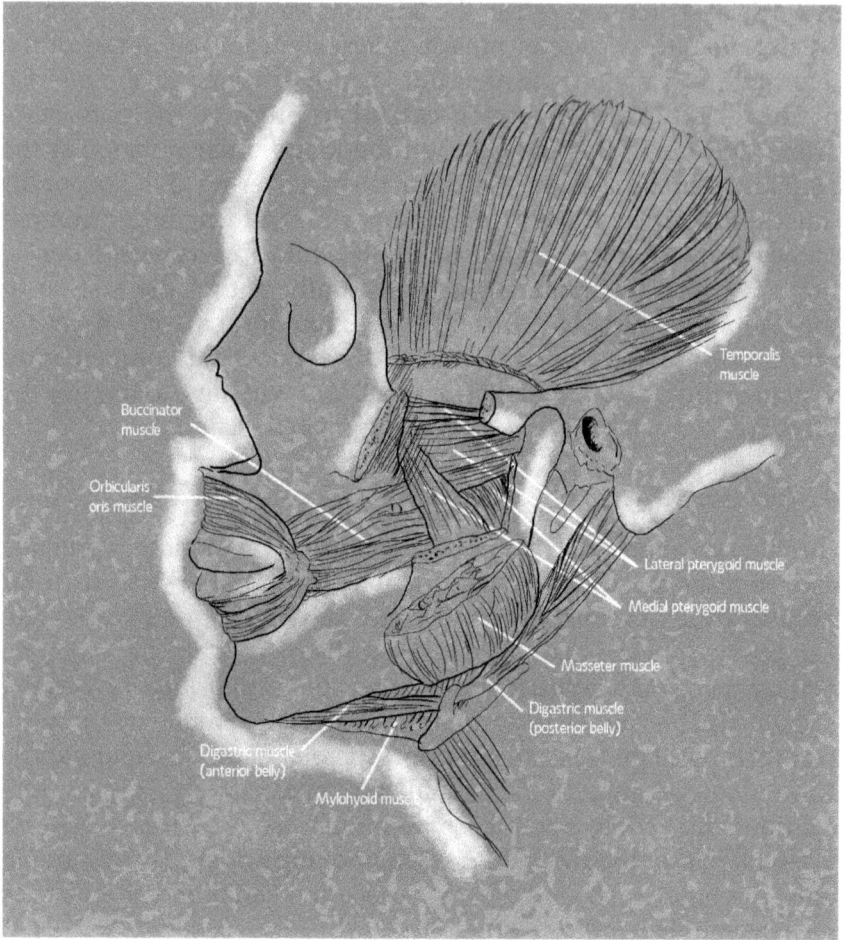

Figure 36. Muscles of mastication.

Lateral (external) Pterygoid muscle

a. Superior head

b. Inferior head

Mandibular foramen

Lingula a small bony tongue-like structure that guards the mandibular foramen and provides attachment for the sphenomandibular ligament.

Nerves

Auriculotemporal runs with superficial temporal artery and it also carries PSNS fibers to the parotid gland.

Masseteric passes through mandibular notch on its way to the masseter.

Inferior Alveolar giving rise to nerve to mylohyoid.

Buccal branch of Trigeminal often attached to tendon of temporalis muscle at the coronoid.

Lingual between mandible and medial pterygoid muscle.

Mylohyoid from inferior alveolar just prior to its entry to mandibular foramen.

Mental termination of the inferior alveolar nerve.

Chorda tympani passes through the petrotympanic fissure to join the lingual nerve 2 centimeters below the cranial vault.

Submandibular ganglion attached to lingual nerve on the surface of the hyoglossus muscle.

Blood Vessels

Superficial temporal artery

Inferior Alveolar artery

Pterygoid plexus of veins on lateral pterygoid muscle, mostly

Maxillary artery (three parts) travels behind the neck of the mandible into the Infratemporal fossa, then passes through the pterygomaxillary fissure to enter the Pterygopalatine Fossa.

For Reference:

FIRST PART

1. Deep Auricular
2. Anterior Tympanic
3. Middle Meningeal–traverses foramen spinosum with nervus spinosus as it passes between the two roots of the auriculotemporal nerve.
4. Inferior Alveolar
5. Accessory Meningeal–may come from maxillary or middle meningeal artery

SECOND PART

1. Masseteric–passes through mandibular notch
2. Pterygoid
3. Deep temporal–may be multiple
4. Buccal

THIRD PART

1. Sphenopalatine——gives nasopalatine branch
2. Descending palatine
3. Infraorbital——accompanies infraorbital nerve thru infraorbital foramen
4. Posterior superior alveolar—supplies upper molars
5. Artery of Pterygoid Canal—accompanies nerve of the same name
6. Pharyngeal
7. Middle superior alveolar
8. Anterior superior alveolar

UNIT 8
Pelvis and Perineum

A. Male and Female Perineum

Male and Female Pelvis (See Figures 37, 38, 39)

1. Preparation

Begin your preparatory work by studying the bones of the pelvis, its bony landmarks, the muscles, and ligaments associated with it in both a male and female pelvis.

2. Cadaver Position

The cadaver may have to be repositioned from supine to prone throughout the dissection and you will need to fully abduct the thighs and tie them in position in order to dissect the perineum. Also, you will need to view a male cadaver if you have a female cadaver and vice versa as the dissection proceeds.

3. Pre-dissection Discussion

A few of the structures that will be uncovered here may have been already uncovered and studied in the gluteal region dissection. A brief overview of the perineum and pelvis will be helpful prior to making the incisions and proceeding with the dissection. The perineum is often described as a "diamond-shaped" area between the femoral regions. In anatomy discussions it is often divided into an Anal Triangle and a Urogenital Triangle by an imaginary line between the anterior portions of the ischial tuberosities. The Anal Triangle is the posterior portion while the Urogenital Triangle is the

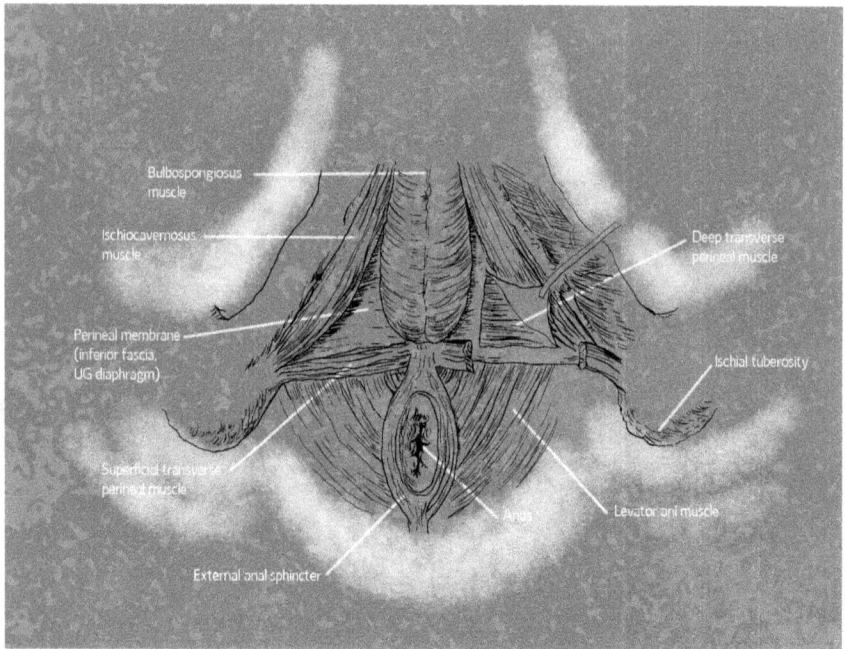

Figure 37. Superficial perineum (male).

anterior portion. In the bony pelvis this "diamond-shaped" area is the inferior pelvic aperture. The anterior point of the perineum is marked by the pubic symphysis, which is covered by the arcuate ligament. The coccyx is the posterior marker. Its side-to-side markers are the ischial tuberosities, while its anterolateral boundary is by the ischiopubic rami. The sacrotuberous ligament provides its posterolateral border. Recall this ligament has the gluteus maximus attaching to it—and you are NEVER to cut it! Begin with dissecting the Ischioanal Fossa, then proceed to sectioning the pelvis only if a prosected specimen cannot be located.

For Reference:

Comparison of Homologous Structures between Male and Female

MALE	FEMALE
Scrotum	Labia Majora
Penis	Clitoris
Corpus Spongiosum	Bulbs of the Vestibule
Glans Penis	Glans Clitoris

Penile Urethra	Urogenital Sinus
Urethral glands	Lesser Vestibular glands
Bulbourethral glands	Greater Vestibular glands
Prostate gland	Paraurethral glands

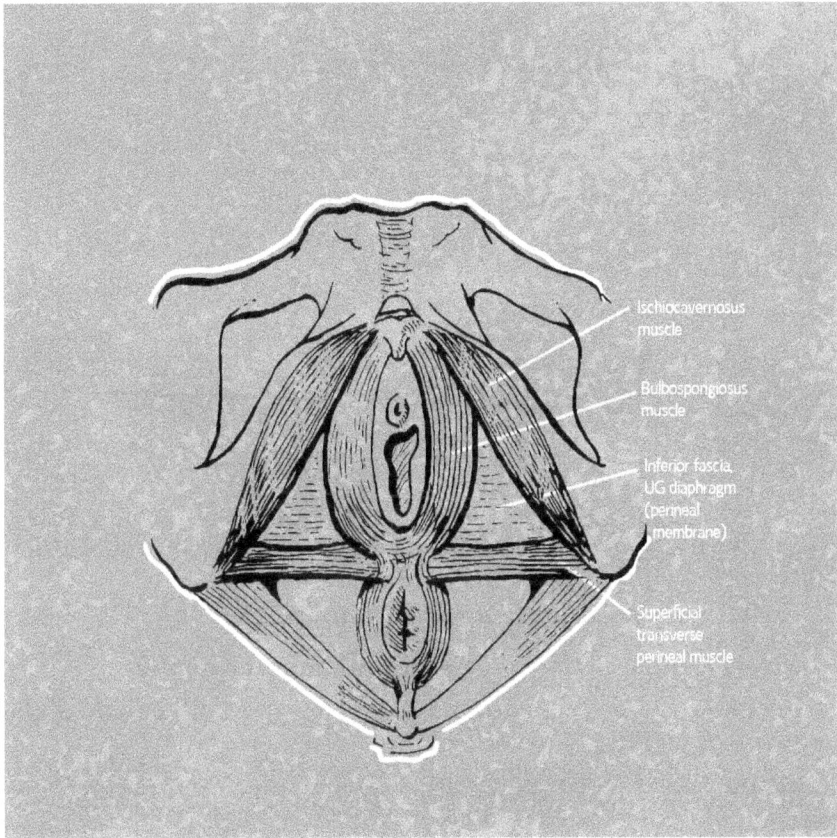

Figure 38. Superficial perineum (female).

4. Incisions to Make (Male)

Open the ischioanal fossa either with a scalpel or puncturing the area with scissors about 1.5 inches deep. Carefully widen the opening. There is usually a fair amount of fat in this fossa that can complicate the dissection.

Mid-sagittal sectioning of the pelvis (both) as well as longitudinal and cross-sectioning of the penis (male) will also be

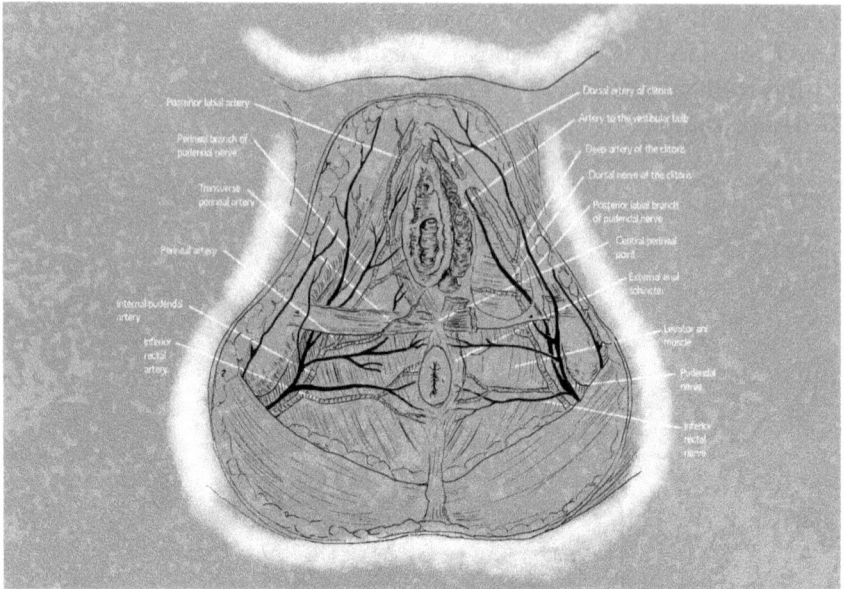

Figure 39. Deep female perineum.

necessary. When sectioning the penis longitudinally place a probe within the urethra and cut on the probe to the prostate and bladder.

5. *Structures to Clean and Identify (Male)*

Muscles, glands, connective tissue and pouches (male)

Ischioanal fossa a wedged-shaped area found on either side of the anus (filled with fat).

Pudendal Canal (Alcock's Canal) a canal in the Obturator Internus fascia.

External Anal Sphincter (three parts)

a. subcutaneous—just encircles the anus.

b. superficial—fixes the anus to perineal body

c. deep—fused to pelvic diaphragm

Fascia of the obturator internus

Superficial Perineal Pouch found between the membranous layer of the superficial fascia (Colle's fascia) and the inferior fascia of the urogenital diaphragm (perineal membrane).

a. superficial transverse perineal—seems to stabilize the perineum
b. bulbospongiosus—covers the bulb of the penis
c. ischiocavernosus—covers each crus
d. crura (legs) of the penis
e. bulb of the penis—vascular erectile tissue

Deep Pouch of the Perineum found between the superior and inferior fascial planes of the deep transverse perineal muscle.

a. deep transverse perineal muscle
b. external urethral sphincter
c. bulbourethral glands
d. branches of internal pudendal artery and pudendal nerve

Perineal body or Central Tendon of the Perineum

a. bulbospongiosus
b. deep transverse perineal
c. external anal sphincter
d. levator ani (iliococcygeus, pubococcygeus and puborectalis)

Penis

a. root
b. shaft
c. glans ("acorn")
d. prepuce (foreskin)
e. frenulum ("rein")
f. external urethral orifice
g. corpus spongiosum
h. corpora cavernosa
i. navicular fossa
j. spongy urethra (the two other parts will be seen later)

Nerves

Pudendal (S2–S4)

   Inferior Rectal

Anterior Scrotal is from the illioinguinal nerve.

Posterior Scrotal is from the perineal nerve which is from the pudendal nerve.

Perineal a terminal branch of the Pudendal as it enters the superficial fascia of the urogenital triangle found below the internal pudendal artery.

Dorsal nerve of the penis a terminal branch of the Pudendal found lateral to dorsal arteries.

Blood Vessels

Internal Pudendal artery is from the anterior division of the internal iliac artery.

Deep artery of the penis anastamose with helicine arteries.

Dorsal artery of the penis termination of internal pudendal artery located on each side of the deep dorsal vein.

Deep dorsal vein of the penis a midline vein that drains into prostatic venous plexus.

6. *Clinical Applications (Male)*

a. **Priapism** a prolonged erection (usually defined as longer than four hours) in the absence of stimulation due to ischemic or non-ischemic causes.

b. **Varicocele** an abnormal dilation of the scrotal venous pampiniform plexus resulting in infertility.

c. **Cryptorchidism** (hidden seed) an undescended testicle which may result in infertility.

7. *Incisions to Make (Female)*

Open the ischioanal fossa either with a scalpel or puncturing the area with scissors about 1.5" deep. Then, carefully widen the opening.

Comparison of Homologous Structures between Male and Female

MALE	FEMALE
Scrotum	Labia Majora
Penis	Clitoris
Corpus Spongiosum	Bulbs of the Vestibule
Glans Penis	Glans Clitoris
Penile Urethra	Urogenital Sinus
Urethral glands	Lesser Vestibular glands "Skene's"
Bulbourethral glands	Greater Vestibular glands "Bartholin's"
Prostate gland	Paraurethral glands

8. Structures to Clean and Identify (Female)

Muscles, glands, connective tissue, and pouches (female)

Ischioanal fossa a wedged-shaped area found on either side of the anus (filled with fat).

Pudendal Canal (Alcock's Canal) a canal in the Obturator Internus fascia.

External Anal Sphincter (three parts)

  a. subcutaneous—just encircles the anus

  b. superficial—fixes the anus to perineal body

  c. deep—fused to pelvic diaphragm

Fascia of the obturator internus

Superficial Perineal Pouch found between the membranous layer of the superficial fascia and the inferior fascia of the urogenital diaphragm (perineal membrane).

  a. superficial transverse perineal

  b. bulbospongiosus

  c. ischiocavernosus

  d. crura (legs) of the clitoris

  e. bulb of the vestibule

  f. greater vestibular glands

Deep Pouch of the Perineum found between the superior and inferior fascial planes of the deep transverse perineal muscle.

  a. deep transverse perineal muscle

  b. external urethral sphincter

  c. urethra

  d. branches of internal pudendal artery and pudendal nerve

Perineal body or Central Tendon of the Perineum attachments

  a. bulbospongiosus

  b. deep transverse perineal

  c. external anal sphincter

  d. levator ani (iliococcygeus, pubococcygeus and puborectalis)

Mons Pubis

  fat pad

Labia Majora
Labia Minora
Clitoris
Prepuce of clitoris
Vaginal Vestibule
External Urethral orifice
Vaginal orifice

Nerves
Pudendal
Inferior Rectal
Perineal
Dorsal nerve of the clitoris
Anterior Labial
Posterior Labial

Blood Vessels
Inferior Rectal
Internal Pudendal
Dorsal artery of the clitoris

B. Male Pelvis

1. Preparation

Begin your preparatory work by studying the bones of the pelvis, its bony landmarks, muscles, ligaments, blood vessels and nerves.

2. Cadaver Position

A prosected specimen can be studied if you cannot create your own sectioned pelvis.

3. Pre-dissection Discussion

See pelvis above and perineum above.

4. Incisions to Make

You will have to reposition the cadaver as you proceed. First make a transverse cut just above the L3 vertebra then section the pelvis (using a saw) in the midline through the pubic symphysis and sacrum. Be sure to section all of the soft tissue and the various organs in their midline also. A scalpel may be used to do the soft tissue cuts.

5. Structures to Clean and Identify (Male) (See Figure 40)

Muscles, glands, organs, connective tissue and pouches (male)
Peritoneum
Perineal membrane
Prostatic urethra
Membranous urethra
Urethral crest
Prostate gland
Seminal colliculus
Prostatic sinus
Prostatic utricle
Openings of the ejaculatory ducts
Ductus (vas) deferens
Rectovesical septum
Seminal vesicle
Ampulla of the ductus deferens
Retropubic space
Retrovesical space
Urinary bladder and trigone
Rectum
    ampulla
    transverse rectal folds
Anal canal
    columns
    valves
    pectinate or dentate line

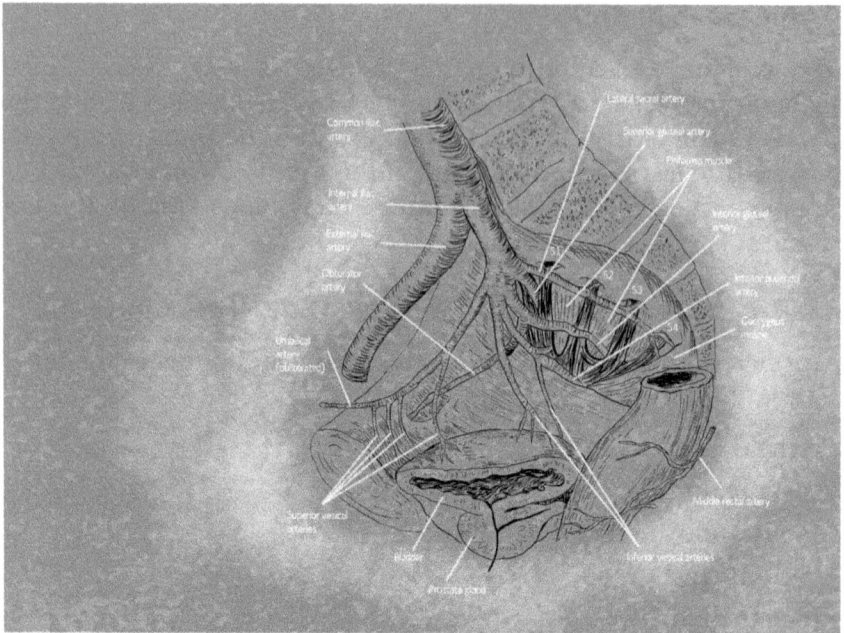

Figure 40. Internal iliac artery branches.

Anal pecten

Nerves
Lumbosacral trunk
Sciatic
S1–S4 ventral primary rami
Pudendal
Pelvic splanchnic nerves (nervi erigentes)

Blood Vessels
Common Iliac artery and vein
External iliac artery and vein
Internal iliac artery (and its branches below) **and vein**
  a. **Obturator artery**
  b. **Umbilical artery**
  c. **Superior vesical arteries** may give rise to deferential artery in the male.
  d. **Inferior vesical artery**

e. **Middle rectal artery** may or may not be present

f. **Internal pudendal artery** passes between the piriform and coccygeus muscles.

g. **Inferior gluteal artery** frequently passes between the S1 and S2 or S2 and S3 nerves and leaves the pelvis just below the piriformis muscle.

h. **Iliolumbar artery** divides into an iliac and lumbar branch.

i. **Lateral sacral artery (s)**

j. **Superior gluteal artery** often located between the lumbosacral trunk and S1.

C. Female Pelvis

1. Preparation

Begin your preparatory work by studying the bones of the pelvis, its bony landmarks, muscles, ligaments, blood vessels and nerves.

2. Cadaver Position

A prosected specimen can be studied, but you may create your own sectioned pelvis.

3. Pre-dissection Discussion

See pelvis above and perineum above.

4. Incisions to Make

You will have to reposition the cadaver as you proceed. Section the pelvis (using a saw) in the midline through the pubic symphysis and sacrum and just above the L3 vertebra. Be sure to section all of the soft tissue and various organs midline also.

5. Structures to Clean and Identify

Muscles, glands, organs, connective tissue and pouches
Peritoneum
Perineal membrane
Vesicouterine pouch
Rectouterine pouch (Pouch of Douglas)

Paravesical fossa

Pararectal fossa

Broad ligament of uterus
  Mesovarium
  Mesosalpinx
  Mesometrium

Parametrium This is found between the two layers of the broad ligament. It is a mass of "areolar" tissue in which lie the uterine vessels, the round ligament of the uterus, the ligament of the ovary and remnants of the mesonephric tubules.

Round ligament of the uterus

Endopelvic fascia

Utero-sacral ligament

Transverse cervical ligament contains the uterine vessels.

Pubocervical ligament

Ovary

Ovarian ligament

Suspensory (infundibular) ligament of the ovary contains ovarian artery and vein

Uterus (parts)
  fundus
  body
  cervix

Uterine tube (parts)
  intramural
  isthmus
  ampulla
  infundibulum
  fimbriae

Vagina

Vaginal fornices
  anterior
  posterior (allows entrance into the Pouch of Douglas for exploratory surgery).
  lateral

Rectum and transverse rectal folds
Urinary bladder and trigone

Nerves
Lumbosacral trunk

Sciatic

S1–S4 ventral primary rami

Pudendal

Pelvic splanchnic nerves (nervi erigentes)

Blood Vessels
Common Iliac artery and vein

External iliac artery and vein

Internal iliac artery (and its branches below) **and vein**

  a. Obturator artery

  b. Uterine artery

  c. Vaginal artery

  d. Umbilical artery

  e. Superior vesical arteries

  f. Middle rectal artery

  g. Internal pudendal artery

  h. Inferior gluteal artery

  i. Iliolumbar artery

  j. Lateral sacral artery

  k. Superior gluteal artery

UNIT 9
Special Dissections

A. Brain Removal

1. *Preparation*

Begin your preparatory work by studying the bones of the skull and their landmarks.

2. *Cadaver Position*

Supine with a block under the thorax and the cervical spine. The cadaver will be turned to prone and then back to supine during this dissection.

3. *Pre-dissection Discussion*

For Reference:

SCALP: The SCALP consists of five layers outlined as following:

Skin: Full of sebaceous glands and hair.

Connective tissue: Dense connective tissue that binds the skin above to the underlying galea aponeurotica. This layer is highly vascular lending to extreme blood loss in even a superficial wound.

Aponeurosis: The galea aponeurotica or epicranial aponeurosis makes up this layer of the scalp. It is a flat tendon connecting two muscles of the scalp, the frontalis, and the occipitalis muscles. These muscles arise near the eyebrows and occiput, respectively, and do not have a direct bony attachment. They insert into the superficial fascia as do most muscles of facial expression.

Loose Connective Tissue: This layer is found just above the Periosteum and immediately below the epicranial aponeurosis. Many deep arteries and veins are found here as well as emissary veins. This is the **"Danger Area of the SCALP"**

Periosteum or pericranium: This layer is fused firmly to bone. It is dense irregular connective tissue.

♦ Removing the brain is easy, but saving all the cranial nerves can be a bit more challenging.

4. *Incisions to Make*

To remove the brain from the cranial vault, the scalp must first be sliced away from the cranium. Cut from the bregma (where the coronal and sagittal sutures meet—you will have to make an educated guess) to the EOP. Then, from the midpoint of this cut, make cuts laterally to the ears. Then, pull the flaps of the scalp downward and laterally to fully expose the bony cranial vault. Tie a string or use a Sharpie to make a cut-line 2.5 centimeters above the upper margin of the orbit and about 1.5 centimeters above the EOP. Be very careful with the bone saw as you cut through the bone into the diploe. Once you are through the diploe, you can "pop" the calvaria off with a t-shaped bone chisel.

Another few more cuts need to be made to remove a chunk of the occipital bone. Place the cadaver in the prone position. Make these cuts almost following what is left of the lambdoid suture to the foramen magnum, avoiding the vertebral arteries. These last cuts will allow you to extract the brain from the cranial vault much easier. You will also need to sever the spinal cord and the vertebral arteries before finally removing the brain from the vault. Return cadaver to the supine position.

You will need to be gentle as you lift frontal lobes and cut the falx cerebri as it attaches to the crista galli of the ethmoid bone. Continue with the "lift and snip" method as you locate and cut the cranial nerves as close to the periosteal dura as you can, saving as much attached to the brainstem as possible.

5. *Structures to Clean and Identify* (See Figure 41)

Inside the cranial vault

Cranial fossae (anterior, middle and posterior)

Crista galli

Cribriform plate

Sella turcica

Hypophyseal fossa

Diaphrama sella

Dorsum sella

Tuberculum sella

Clinoid processes (anterior, posterior, and middle)

Cavernous sinus and its contents

Pituitary gland

Sphenoidal jugum

Trigeminal ganglion

Structures or spaces associated with the brain after its removal (See Figures 42 and 43)

Cranial nerves I–XII

Dura mater
>   Periosteal layer
>   Meningeal layer

Arachnoid mater

Subarachnoid space

Dural venous sinuses

Falx cerebri

Falx cerebelli

Tentorium cerebelli

Grooves for middle meningeal artery on the frontal and parietal bones

Vertebral arteries

Basilar artery

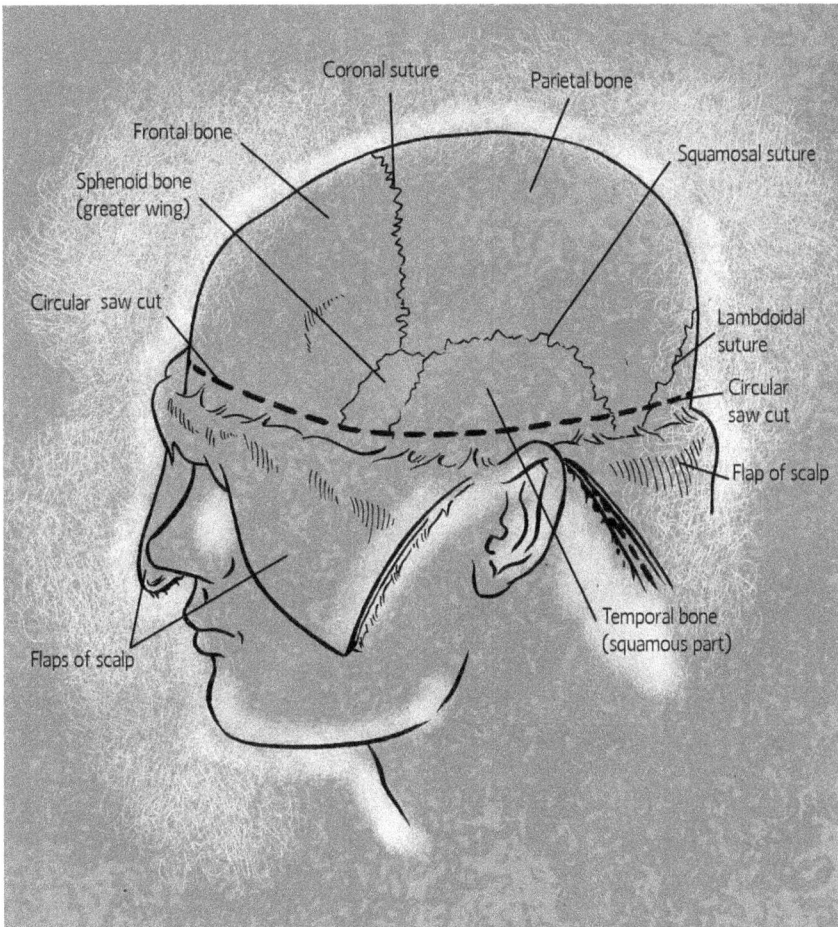

Figure 41. Incisions to remove the scalp.

Circle of Willis (and its formation)
Vertebral arteries
Basilar artery
Labyrinthine artery
Pontine branches
Posterior inferior cerebellar (PICA)
Anterior inferior cerebellar
Superior cerebellar
Posterior cerebral (PCA)

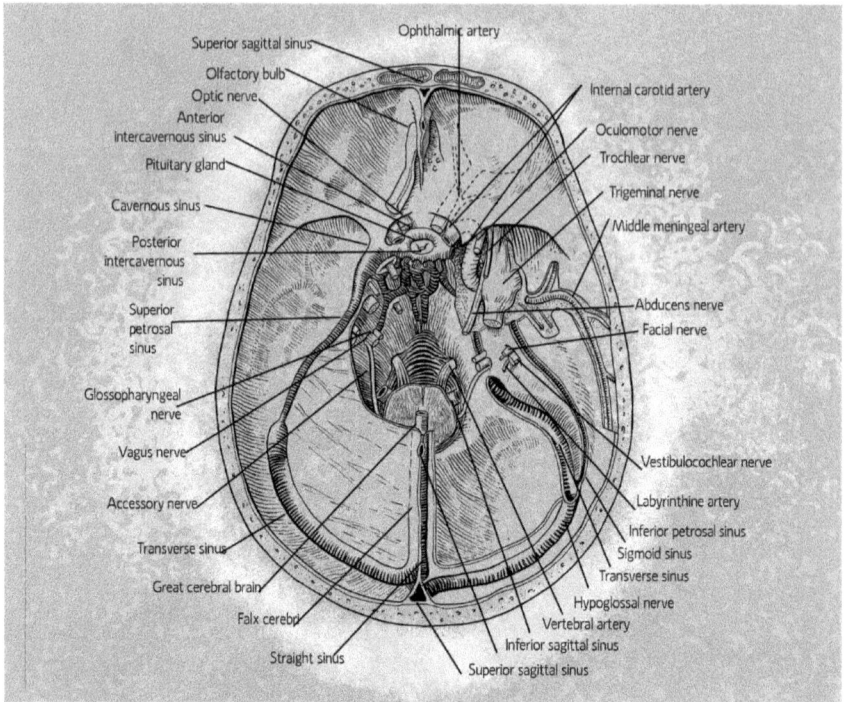

Figure 42. Superior view inside the cranial vault.

Middle cerebral (MCA)
Internal carotid artery (ICA)
Anterior cerebral artery
Anterior communicating
Anterior spinal artery
Posterior spinal arteries
Cerebellopontine angle
Medulla
Pons
Midbrain (tectum and tegmentum)
Superior and Inferior colliculi
Ventricles and cerebral aqueduct

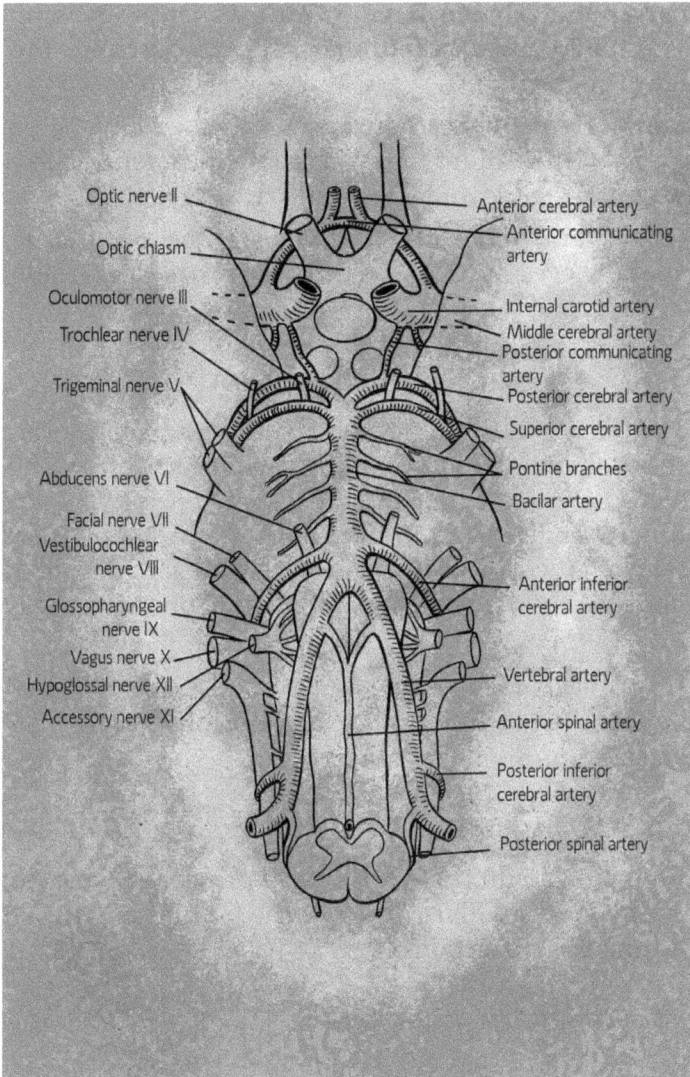

Figure 43. Ventral view of medulla and pons showing arteries of Circle of Willis.

6. Clinical Applications

a. **Meningitis** inflammation of the meninges of either bacterial or viral origin. May be divided into pachymeningitis (dura only) or leptomeningitis (arachnoid and pia together).

b. **Plagiocephaly** an asymmetric flattening of the head either due to synostotic causes or deformational changes.

c. **Subarachnoid Hemorrhage** rupture of an artery within the subarachnoid space resulting as patient's describe "the worst headache of my life."

d. **Epidural Hematoma** an arterial rupture within the epidural space often from blunt force trauma to the lateral aspect of the skull rupturing the middle meningeal artery.

e. **Racoon Sign (periorbital ecchymosis)** the pooling of blood around the eyes due to a fractured skull base, especially of the anterior cranial fossa.

B. Disarticulation

Prior to sectioning the head for study of the nasal cavity, oral cavity, and paranasal sinuses, it will need to be disarticulated from the body. If you have access to more than one cadaver remove its head too, but only as described in the Pharynx and Larynx portion of this dissection. If you do not have access, then a model will have to suffice for study of the posterior and internal pharynx and larynx.

1. Preparation

Begin your preparatory work by studying the landmarks of the skull, nasal cavity, and associated viscera.

2. Cadaver Position

Prone with a block under the thorax to provide for anterior neck flexion. The cadaver will need to be turned supine and then back to prone during this dissection.

3. Pre-dissection Discussion

This dissection and subsequent study of the area involves severing attachments and preserving important structures in the region. Making a clean separation of the head from the torso is imperative. In doing so you will be able to clearly identify the structures of interest.

4. Incisions to Make

Anteriorly, decide where to make the transverse cut across the cervical spine. Typically, it is done between C7 and T1 vertebrae.

Posteriorly, cut all of the muscles attaching to the skull. It is up to you how much of the muscle you wish to save. Sever the spinal cord, vertebral arteries, and all the associated nerve structures. If you have not cut the sympathetic chain, do it now. The head should now be detached from the torso.

C. Bi-sectioning of the Head

1. Preparation

The head was detached previously and now is free to position for bi-sectioning. It is useful to have a bandsaw to do this, but a handsaw also will work well.

2. Cadaver (head) Position

The head is to be placed face up on a solid surface. Make sure that the head does not slip or roll at all. It needs to be perfectly still and anchored so that the proper cut can be made.

3. Pre-dissection Discussion

It is important to identify (if you can) the nasal septum first. You will cut just lateral to it, not midline. Doing so will preserve the septum and allow two different presentations of the nasal cavity. You may also need to realign the soft tissue or various cartilaginous structures to section them properly.

4. Incisions to Make

If using a handsaw, place the saw blade at (or just lateral to) the internasal suture. Saw through the nasal bone, frontal bone, sphenoid bone, ethmoid bone, nose, hard and soft palates. Do your best to section the tongue, lips, and tracheal cartilage in the midline. Continue the cut through the foramen magnum and cervical vertebrae. You may need to use a scalpel here and there to clean up certain areas.

5. Structures to Clean and Identify

Muscles, glands, and spaces
Nasopharynx
Oropharynx

Laryngopharynx

Frontal sinus

Septal cartilage

Nasal vestibule

Superior, middle, and inferior nasal conchae

Superior, middle, and inferior meatuses

Ethmoidal bulla

Tongue

Sulcus terminalis

Lingual frenulum

Plica fimbriata

Uvula

Lingual tonsils

Foramen cecum

Papillae
    fungiform
    circumvallate
    filiform
    foliate

Median glossoepiglottic fold

Valleculae

Lateral glossoepiglottic fold

Piriform recess

Hyoglossus muscle

Genioglossus muscle

Mylohyoid muscle

Stylohyoid muscle

Geniohyoid muscle

Anterior belly of digastric

Styloglossus muscle

Palatoglossal fold

Palatopharyngeal fold

Isthmus of the fauces

Palatine tonsil

Submandibular gland

Submandibular duct found in the floor of mouth superficial to the lingual nerve.

Sublingual glands and ducts

Tensor veli palatini

Levator veli palatini

Sphenoethmoidal recess

Hiatus semilunaris

Sphenoidal sinus

Maxillary sinus

Openings for (frontal sinus, anterior ethmoidal cells, maxillary sinus, middle ethmoidal cells, posterior ethmoidal cells, sphenoidal sinus).

Opening of pharyngotympanic tube

Salpingopharyngeal fold

Salpingopharyngeus

Torus tubarius

Pharyngeal tonsil

Nerves

Lingual in the floor of the mouth traveling under the submandibular duct.

Hypoglossal located at the inferior border of hyoglossus muscle between it and the submandibular gland.

Submandibular ganglion found on the surface of the hyoglossus muscle.

Blood vessels

Lingual artery

D. Separation of the Head and Cervical Viscera from the Vertebral Column

1. *Preparation*

Be aware of the contents and locations of structures in this region. Know exactly why you are doing this particular dissection and what structures you are interested in preserving.

2. Cadaver Position

Place the cadaver supine with the head in mild extension.

3. Pre-dissection Discussion

You will need some strength to do this particular separation.

4. Incisions to Make

You will need to make a transverse cut across the trachea and esophagus at approximately the C6 vertebra. Also, you will need to cut the neurovascular structures in the area too. Be sure to save some length of them so they can be easily studied once the head is removed. After you have made those cuts, place your hand deep to the esophagus in the prevertebral space and lift them from the prevertebral fascia and deep cervical musculature. Once it is loose, flip the cadaver over and cut any structures that are holding the head in place. Next, place a chisel in the atlanto-occipital joint and disarticulate the skull from the vertebral column. Flip the cadaver back over and tug gently to remove the head and cervical viscera. You may need to cut any remaining muscles that are holding it tight.

5. Structures to Clean and Identify: Posterior Pharynx (See Figure 44)

Superior pharyngeal constrictor
Middle pharyngeal constrictor
Inferior pharyngeal constrictor
Cricopharyngeus
Buccopharyngeal fascia
Retropharyngeal Space
Styloglossus muscle
Palatoglossus muscle
Stylopharyngeus
Pharyngeal raphe
Pterygomandibular raphe
Pharyngobasilar fascia
Vagus nerve

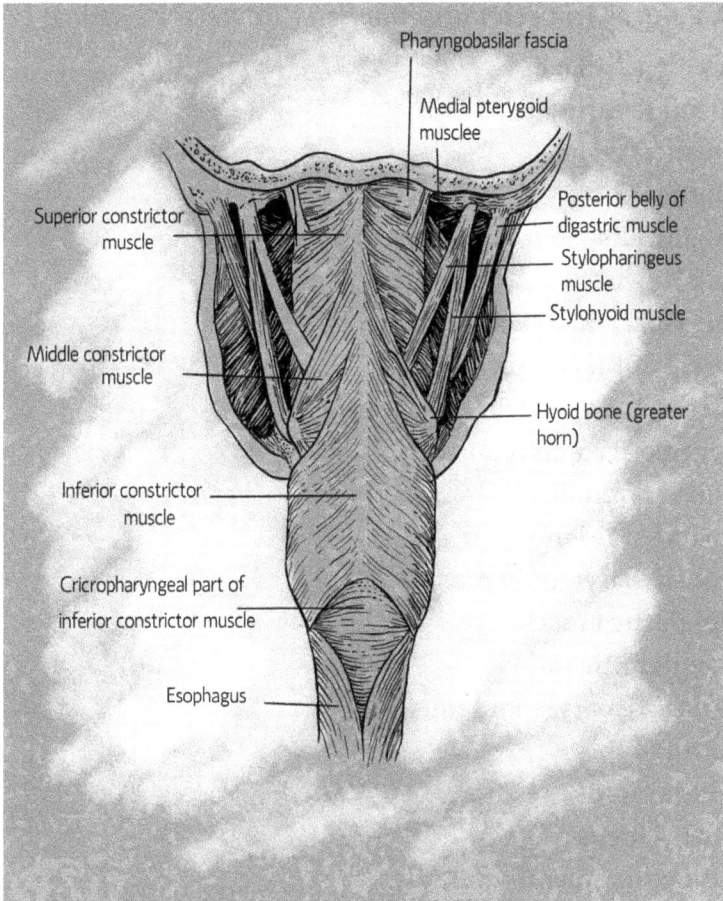

Figure 44. Posterior view of pharynx.

Superior laryngeal nerve
Internal laryngeal nerve
External laryngeal nerve
Spinal accessory nerve
Sympathetic chain with superior cervical ganglion
Hypoglossal nerve
Glossopharyngeal nerve

6. *Structures to Clean and Identify: Posterior Larynx*

Arytenoid cartilages

Corniculate cartilages

Triciteal cartilages

Epiglottis

Rima glottidis--space between true vocal ligaments

Valleculae

Glottis (space with vocal ligaments)

Ventricle of the larynx

Saccule of the ventricle

Transverse arytenoid muscle

Oblique arytenoids

Thyroarytenoid muscles

Lateral cricoarytenoid muscle

Aryepiglottic muscle

Thyroepiglottic muscle

Posterior cricoarytenoid muscle

Vocal folds

Vestibular folds

Conus Elasticus

Quadrangular membrane

E. Orbit

1. *Preparation*

Doing this dissection on a head that has not been disarticulated with be easiest. Study the bones that form the orbit and its contents. Pay special attention to their location within the orbital cavity. Be aware that the structures in the eye are very delicate and extra caution is needed in order to preserve them.

2. *Cadaver Position*

Supine with a block under the head and thorax. Be certain the head is in mild flexion for easy access to the orbit.

3. *Pre-dissection Discussion*

You will be entering the orbital cavity from the superior aspect. You will need to remove the periosteal dura from the floor of the anterior cranial fossa.

The orbital plate is delicate, but a chisel and hammer will be needed. Go slowly, but firmly and be careful to preserve the structures of interest. There is a lot of periorbital fat here and it needs to be carefully removed with forceps to uncover the structures of interest.

4. *Incisions to Make*

About 2 centimeters lateral to the crista galli gently chisel through the orbital plate. Chisel and pick through the pieces of the orbital plate. No cuts, just chisel and pick. A tweezers or forceps may be needed to pick the pieces out. By continuing to chisel and pick out the pieces you will create a large circular opening that looks directly into the superior aspect of the orbit.

5. *Structures to Clean and Identify* (See Figure 45)

Muscles

Levator palpebrae superioris

All extraocular muscles
    Superior oblique with its trochlea
    Lateral rectus
    Medial rectus
    Inferior oblique
    Inferior rectus
    Superior rectus

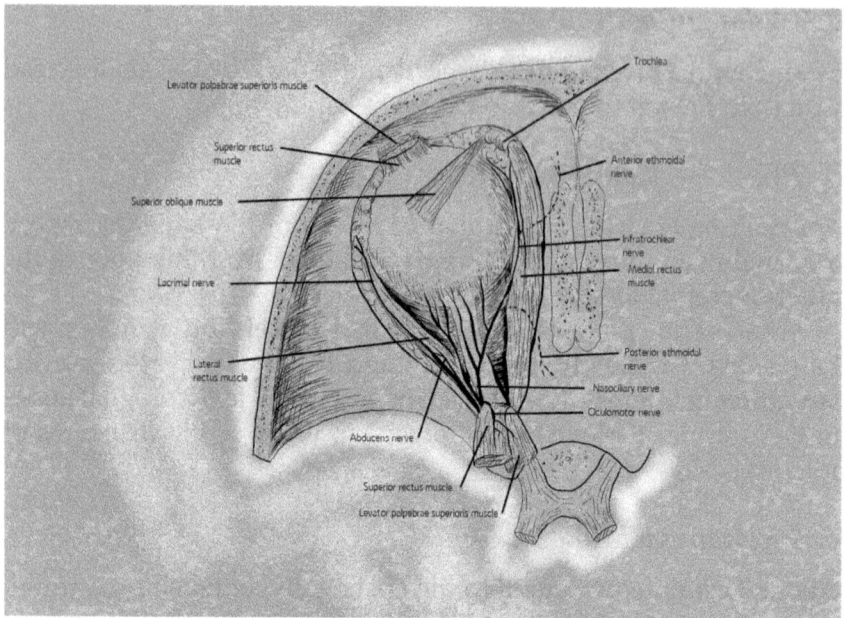

Figure 45. Superior view of left orbit.

Nerves
Frontal
    Supraorbital
    Supratrochlear

Nasociliary

Lacrimal

Oculomotor

Abducens

Trochlear

Optic nerve

Ciliary ganglion

Index

For Product Safety Concerns and Information please contact our EU
representative GPSR@taylorandfrancis.com
Taylor & Francis Verlag GmbH, Kaufingerstraße 24, 80331 München, Germany

www.ingramcontent.com/pod-product-compliance
Lightning Source LLC
Chambersburg PA
CBHW070727220326
41598CB00024BA/3332